The Metaphysics of Michael Polanyi

Martin E. Turkis II

The Metaphysics of Michael Polanyi

Toward a Post-Critical Platonism

palgrave
macmillan

Martin E. Turkis II
San Francisco, CA, USA

ISBN 978-3-031-41210-3 ISBN 978-3-031-41211-0 (eBook)
https://doi.org/10.1007/978-3-031-41211-0

This Palgrave Macmillan imprint is published by the registered company Springer Nature Switzerland AG.
The registered company address is: Gewerbestrasse 11, 6330 Cham, Switzerland

Paper in this product is recyclable.

For Benji and Fitzy

Preface

In this book I tell the story of how the Platonic metaphysical vision of Michael Polanyi—a Hungarian-British chemist and philosopher—can bridge the gap between speculative metaphysics and scientific practice, thus making sense of the broad swathe of human experience in a phenomenologically satisfying fashion.

Polanyi is best known for his post-critical epistemology which posits necessary personal and tacit coefficients as fundamentals of all knowing, and I therefore propose to tell this tale through the investigation and development of the metaphysics implied by his epistemological work.

I develop key concepts from Polanyi's theory of knowledge—especially the *comprehensive entity*—explain how his Platonist metaphysics fit within his understanding of the tradition of Copernicus, show how Polanyi is a Platonist with respect to universals, and sketch out the tripartite contours of a post-critical Platonism: an Eleatic metaphysical hierarchy tracked by causal potency, a flat ontology of comprehensive entities, and a mereology that explains the relationship of comprehensive entities to the subsidiary particulars of which they are constituted.

Existing scholarship on Polanyi focuses primarily on the nuances of his theory of knowledge and secondarily on its application in a range of fields, from science to theology to business. The metaphysical implications of his work, however, while touched on from time to time, have not been explored with the same thoroughness or systematic rigor. This is a notable lack given that Polanyi himself claims that the motivation for formulating his epistemology is the justification of the "metaphysical grounds of intellectual life in a free, dynamic society" (*TD* 83). My work thus seeks to fill a significant gap in the literature on Polanyi's philosophical contributions by offering up the first in-depth exploration of Polanyi's metaphysics.

San Francisco, CA, USA Martin E. Turkis II

Acknowledgments

Many friends, colleagues, and mentors have played indispensable roles in the completion of this research project. First and foremost I wish to extend my gratitude to Paloma Pérez-Ilzarbe and Francisco Gallardo de la Torre, who patiently yet critically directed and co-directed my Ph.D. dissertation on Polanyi's metaphysics at the University of Navarra. Through their participation in my dissertation committee, Jaime Nubiola, Gonzalo Génova, and Nathaniel Barret were immensely helpful in refining my project, as were Walt Gulick and Dale Cannon, stalwart members of *The Polanyi Society* whose works I cite extensively herein.

I have also profited from the invaluable feedback and support of many other *Polanyi Society* members. Phil Mullins has tirelessly read and offered compelling and constructive critiques of nearly everything I present here. Through their engagement with various drafts and papers, Jon Fennell, Richard Moody, Andrew Grosso, and Esther Meek have all helped to refine my arguments as well, as has Charles Lowney, who worked tirelessly and lent his considerable expertise to the refinement of this work for publication. I have also benefited from the in-depth conversations facilitated by Stan Scott, organizer of the Thresholds Group, which included Jean Bocharova, Eric Rustad, Dale Cannon, and myself. These intimate meetings were incredibly helpful in working out the contours of the proposals contained herein.

Geoffrey Green's guidance, both while writing my earlier M.A. thesis on English Literature and later when shifting my focus to philosophy, was fundamental. And conversations with my close friend and colleague Erik Migdail have been an indispensable source of insight and wisdom throughout the process.

I am thankful also for the love and support of my wife, Grace, and my two sons, Fitzy and Benjamin. And I am deeply grateful to my parents, Martin and Deborah Turkis, for all of their support, from childhood through the present.

Contents

About the Author

Martin E. Turkis II is a teacher, musician, and philosopher residing in San Francisco. He received his Ph.D. in Philosophy from the University of Navarra, Pamplona, Spain, for his research on the metaphysics of Michael Polanyi and Plato. His other interests include political economy, virtue ethics, and philosophy of education. He is the editor of *Tradition and Discovery*, the journal published by *The Polanyi Society*.

Abbreviations[1]

CI	Creative Imagination
FR	Faith and Reason
KB	Knowing and Being
M	Meaning
PK	Personal Knowledge
SFS	Science, Faith, and Society
SM	Study of Man
TD	Tacit Dimension

[1] Works by Michael Polanyi, which are frequently cited, will be abbreviated as follows.

Introduction

It will come as no surprise to readers of this volume that since the dawn of the modern era metaphysics has generally had a rough go of it. The big questions that have tended to define metaphysics in the Western tradition were supplanted – in terms of scholarly respectability, at least – by epistemology and other sub-disciplines that in pre-modern thought might have been construed as offshoots of the more fundamental work of metaphysical speculation.

Nevertheless, it has often been noted that metaphysical issues continued to tacitly permeate philosophical debates. This, of course, could not be otherwise. Thus toward the end of the twentieth century and on into the 21st we have seen a resurgence of interest in doing metaphysics openly and explicitly in both the analytic and continental (or post-continental) spheres, a phenomenon that has given rise to a rich variety of theories including but not limited to speculative realism (including materialist and anti-materialist variants), neo-Aristotelianism and contemporary hylomorphism, structural realism, neo-Humeanism, etc. This rebirth certainly has its share of critics who long for the days when metaphysics

M. E. Turkis II, *The Metaphysics of Michael Polanyi*,
https://doi.org/10.1007/978-3-031-41211-0_1

was an object of open derision, and one can find contemporary metaphysicians who brand their distinctive approaches with novel terminology or by stipulating definitions of the field that are far more restrictive than more traditional understandings of the discipline have tended to be. There is, in short, no end of disagreement among the participants in the renewed debates, but there can be no doubt that metaphysics has returned, and with a vengeance.

I thus propose to investigate the metaphysics of Michael Polanyi. Polanyi was a Hungarian chemist turned economist and philosopher who is best known as an epistemologist, a status which is justly due to his most influential contribution on the philosophical scene, the theory of tacit knowing. Over the course of his philosophical career Polanyi systematized his epistemology quite thoroughly, though he never focused on or developed to any significant extent the metaphysics implied by this epistemology. Nonetheless he often enough stated explicitly that the aim of his theory of knowledge was to restore our ability, largely lost during what he called the critical revolution in Western thought, to make metaphysical claims in good faith.

The goal for this project, then, is to draw out the metaphysics tacitly at work in Polanyi's post-critical[1] project, make it explicit, and then to develop it where necessary, thus bringing it closer to the sort of systematic fruition characteristic of Polanyi's epistemology. The point of departure within Polanyi's corpus is what I call his Copernican Realism, a metaphysical realism that affirms both abstract philosophical and concrete empirical approaches to reality as valid. Beyond this, I will argue that a Polanyian, post-critical metaphysics is, in essence, Platonist in its explicit top-downism (Gerson 2005) as well as its embrace of the full reality of abstract or immaterial entities. Such a post-critical Platonism, however – and despite the knee-jerk reactions that this claim will doubtless inspire – need not imply a reified Two Worlds Theory involving a Platonic Heaven any more than an acknowledgement that the game of chess is governed by the rules of chess implies the existence of a Chess Rule Heaven.

[1] Chapter 1 clarifies Polanyi's use of the term "post-critical." For the present it is sufficient to note that the term does not simply refer to post-Kantian thought but rather to Polanyi's recommended fiduciary approach to philosophy.

The post-critical Platonist metaphysics developed here will, however, insist that *to be* is *to mean* and will involve a metaphysical hierarchy defined by the relative causal power of different entities to effect change in the world around them, or, as Polanyi would say, to indefinitely manifest themselves in the future (i.e. the Eleatic Principle or the IFM Effect).[2] Within this hierarchical, Eleatic metaphysic operates a flat ontology of comprehensive entities characterized by mereological relations between subsidiary particulars and the integrative joint meanings that define the comprehensive entities themselves. Such a succinct statement of a relatively sweeping set of metaphysical claims will of necessity be cryptic at this point; all shall be made clear as the project proceeds.

The structure of the book is as follows. The following chapter begins with a brief biography of Michael Polanyi for those less familiar with his life and work. This will set the stage for an examination of the post-critical approach to philosophy, the larger project towards which his epistemological work was oriented and which provides the framework within which my own project unfolds. This chapter includes a section summing up Polanyi's theory of tacit knowing and will also make some preliminary gestures towards a rapprochement between Polanyi and Plato, especially with regard to the ways that the legacies of both make room for *aporia*.

From there we will move on to an important element of Polanyi's metaphysics about which he was relatively explicit – his metaphysical realism. Based on a close reading of a series of lectures given by Polanyi later in life which culminates in the article "Science and Reality" (1967), I classify Polanyi's as a Copernican realism – that is to say, a realism that embraces both philosophy of first principles *and* empirical research as valid means of grasping reality. I then attempt to bring his Copernican realism into conversation with three contemporary metaphysical realisms: Ontic Structural Realism, Object-Oriented Ontology, and Neo-Aristotelianism.

We then follow Polanyi as he moves from epistemology to metaphysics by means of what he calls the ontological aspect of tacit knowing. This sets us up for a deep dive into further metaphysical topics which are implied, though never systematically or explicitly fleshed out in Polanyi's

---

[2] See Plato's *Sophist* 247d-e (1997) and Esther Meek (2017, 77).

corpus. Most importantly we take a close look at Polanyi's understanding of what he called *comprehensive entities*, the joint meanings reached epistemologically by means of the tacit integration of participating subsidiary particulars. Polanyi's arguments for the ontological aspect imply that the structure of comprehensive entities is ontological as well as epistemological. I develop this Polanyian notion further by arguing for a flat ontology of comprehensive entities, which is tantamount to the claim that all existents are characterized by this subsidiary particular/joint meaning mereological structure. Another conclusion reached on the basis of this chapter's exploration is that to be (or to exist) is to mean, which is to say that reality is potentially intelligible through and through, though this does not imply any sort of mind-dependence. This potential intelligibility is rooted in Polanyi's understanding of indwelling and the tacit dimension of knowledge.

The following chapter examines the roles of the material and the immaterial in a post-critical Platonist metaphysics. It more explicitly connects the Polanyian metaphysical notion of comprehensive entities to Platonist metaphysics, primarily by arguing that Polanyi's controversial field theory implies the existence of Platonic Forms through the full ontological status it accords to potentiality. This leads to an examination of Polanyi's explicit teachings on universals, since he, like many others, takes Forms to be universals (or rather a particular theoretical approach to universals), though this is arguably incorrect. We further blend Polanyi's post-critical metaphysics with Platonism by arguing that as existents, Forms are comprehensive entities. In addition, I introduce a subcategory that I call the Primordial Forms which includes those Forms which are necessarily instantiated in all possible worlds in which there is a physical stratum of reality.

In light of the foregoing topics of discussion, we move next to a comparison of Polanyi, Plato, and Aristotle's respective theories of concept acquisition. On the basis of close readings of selections from *Posterior Analytics* and *Phaedo*, I argue that Aristotle's theory is insufficient if we take universals to be metaphysical rivals to Plato's Forms. I also examine, to the extent possible, Polanyi's misunderstanding of the theory of Forms and then argue for an unacknowledged sympathy between Polanyi's teachings on universals and Plato's understanding of concept acquisition that allows us to leverage the tacit dimension of knowing toward a reconception of *anamnesis*, a move that Polanyi himself began to explore in his analysis of the *Meno.*

The seventh chapter further develops the post-critical Platonism that takes shape as our examination unfolds, situating it within Plato studies as well as the widespread simplifications and misunderstandings of Plato's work which predominate outside the context of scholarship on Plato. Additionally, there is further working out of our understanding of Forms as comprehensive entities and arguments in favor of taking Forms and universals to be distinct types of entities (rather than semi-synonymous terms or rival metaphysical theories), both with full ontological status, although Forms occupy a higher rung of the metaphysical hierarchy.

1 A Note on Polanyi's Integrative Approach

Polanyi's post-critical approach to philosophical questions is fundamentally integrative. As we will see in the next chapter's review of his epistemology, subsidiary details are tacitly integrated into comprehensive entities or joint meanings. According to his theory of tacit knowing, attending focally to a subsidiary component shatters our comprehension of the joint meaning to which it contributes, at least for as long as we sustain this focal attention to the subsidiary. In light of this it ought to be unsurprising that Polanyi often approaches philosophical issues without reference to categorical distinctions common in analytic philosophy. One example that exemplifies this quite clearly would be the general metaphysical realism that Polanyi claims is assumed, tacitly or explicitly, by working scientists. This, as we will see in the third chapter, is a position which includes within its ambit what many analytic thinkers would categorize as scientific and epistemological realisms. Polanyi, however, does not tend to treat the issues taken up under these latter two headings as separate from his overall project but rather handles them as various facets of a larger, comprehensive whole, the joint meaning to which they contribute. In this general tendency or orientation, I follow his example.

References

Gerson, Lloyd P. 2005. *Aristotle and Other Platonists.* Ithaca, NY: Cornell University Press.

Meek, Esther Lightcap. 2017. *Contact with Reality: Michael Polanyi's Realism and Why It Matters.* Eugene, OR: Cascade Books.

Plato. 1997. *Sophist* trans. Nicholas P. White, in *Plato: Complete Works.* Edited by John M. Cooper. Indianapolis, IN: Hackett Publishing Company.

Polanyi, Michael. 1967. "Science and Reality." *The British Journal for the Philosophy of Science* 18/3: 177–196.

Michael Polanyi and the Post-Critical Approach to Philosophy

1 Biographical Details

1.1 Early Life[1]

For those who are unfamiliar, Michael Polanyi (originally Mihály Pollacsek, later Magyarized to Polányi) was a Hungarian-British polymath who made significant contributions in the areas of physical chemistry, economics, and philosophy. He was born in Budapest in 1891, and died in 1976 in Northampton, England.

He grew up in an assimilated, secular and cosmopolitan Jewish family which was quite wealthy, though they suffered significant economic setbacks: First, a declaration of bankruptcy in 1900 when a natural disaster caused the collapse of a railway project Polanyi's father was building; and later, in 1905, when he (Polanyi's father) passed away.

[1] In order to avoid constant citations in this non-technical chapter I will direct interested readers to the appropriate chapters of my biographical source texts. Topics covered in the section entitled "Early Life" can be studied in greater detail by consulting the first two chapters of Scott and Moleski's *Michael Polanyi: Scientist and Philosopher* (2005) as well as Chap. 1 of Mary Jo Nye's *Michael Polanyi and His Generation: Origins of the Social Construction of Science* (2011).

M. E. Turkis II, *The Metaphysics of Michael Polanyi*,
https://doi.org/10.1007/978-3-031-41211-0_2

At home the children spoke German with their mother, Cecile, and also learned English and French. The family culture was one of intellectual vanguardism – Cecile held regular salons at which artists, writers, and playwrights debated politics, current events, and aesthetics. There was a significant dose of left-wing politics in Polanyi's family (most pointedly among his siblings and mother), and as a young man he was a member of a left-leaning student group, the Galileo Circle, although he ultimately became a moderate liberal.

Polanyi began studying to be a physician in 1908. During his medical studies he began to publish scientific papers. He received his medical degree in 1913 and relatively soon thereafter was serving as a medical doctor with the Austro-Hungarian forces in World War I. His health deteriorated in 1914 and he was sent back to Budapest on furlough to recover. During the rest of the war he performed light duties and found time to carry out scientific research and to complete a thesis for a Ph.D. in physical chemistry at the University of Budapest.

After the war, Polanyi, like his brother Karl, served in the government of the newly formed Republic of Hungary. He resigned immediately when in 1919 Bela Kun effected his communist takeover of the government and declared the short-lived Hungarian Soviet Republic, soon to be replaced by Horthy's Christian and ethnically nationalist regime, an option equally repugnant to Polanyi. Under pressure from increasingly anti-Semitic policies under Horthy, many nonobservant Jews converted, often nominally, to various Christian denominations.

Polanyi himself was baptized into the Catholic Church in October of 1919, soon before he left Hungary for Germany. His conversion, however, was not one of mere expediency, since he considered himself around this time a "converted Christian on the lines of Tolstoy's confession of faith," a theological stance similar to his brother, Karl's, although Karl did not convert to Catholism (Nye 2011, 10; quoted from a letter to Karl Mannheim dated April 14, 1944).

1.2 Germany[2]

Polanyi moved to Karlsruhe, Germany in 1919. It was in Germany that he met and married another Hungarian emigré, Magda Kemeny in 1923 and they remained in Germany until 1933, when, under the looming threat of Nazi political power, they left for Manchester, England. Thus Polanyi became a part of what has been called the "Hungarian Phenomenon," a twice-exiled group of gifted scientists and mathematicians including von Neumann, Eugene Wigner, Szilard, and others who left Hungary to escape the harsh Horthy regime and were later forced to relocate once more.

In 1920 he moved to Berlin after accepting a job at the Kaiser Wilhelm Institute for Fiber Chemistry, where he worked until he transferred in 1923 to the Institute for Physical Chemistry (later the Fritz Haber Institute), where he remained until leaving Germany. This was a rich period of scientific and intellectual camaraderie and chemical research for Polanyi. He became one of the world's leading experts on adsorption of gases, worked widely in the areas of chemical kinetics and x-ray diffraction, and developed theories of fiber diffraction analysis as well as of plastic deformation of ductile metals. Much cutting-edge research was going on in Germany at the time, and there was correspondingly a tightly-knit, collegial scientific culture with which Polanyi was involved. He participated regularly, for example, in a Wednesday colloquium organized by physicist Max von Laue, where "Schrodinger…Planck, Nernst, Einstein, and Haber could often be found" (Nye 2011, 50). Polanyi later recalled these colloquia as one of the high points of his intellectual life (Scott and Moleski 2005, 85). He explored intellectual interests outside the physical sciences as well, founding an economics dining club which was regularly attended by friends such as Wigner, von Neumann, and others.

Eventually the rise of Nazi power and institutionalized anti-Semitism forced a reluctant Polanyi to move his family and his research to England, where he taught and did research in the Chemistry Department at the University of Manchester. Nevertheless, this time in Germany was critical in the formation of Polanyi's later philosophical ideas. As Nye puts it,

[2] See Scott and Moleski (2005) Chaps. 3, 4, 5, and Nye (2011) Chaps. 1, 2, 3.

> Michael Polanyi's thirteen years in Berlin gave him the experience of what he later transformed into an idealized vision of the scientific research community. Polanyi's later reflection on the world he had lost propelled him into an intellectual exposition and defense of the social and institutional preconditions that are necessary for the flourishing of modern science…Like the social and institutional arrangements that were the fabric of his scientific life, the daily routines of doing science were also the stuff of which his philosophy of science was to be made. (2011, 83–84)

1.3 Manchester and the Transition to Philosophy and the Social Sciences[3]

The move to Manchester took place in 1933. Polanyi continued his research and taught classes on chemistry and quantum mechanics, though his interests wandered further and further from the sciences, drifting more to the politics and economics that had always fascinated him, in part because he feared that a lack of economic sophistication on the part of the public contributed to anti-Semitism. Eventually he was doing relatively little chemistry, and in 1948 the university would move him from the Department of Chemistry to Social Sciences, where a new chair was created for him.

Politically, Polanyi tended toward a nineteenth century style of cosmopolitan liberalism, and he was a member of the Mont Pelerin Society.[4] Yet while his economic views tended correspondingly towards the free markets of classical liberalism, he became a moderate Keynesian upon reading Keynes's *General Theory* in 1936, much to the annoyance of his friend F.A. Hayek and others from the Mont Pelerin Society, which he eventually abandoned as out of accord with his vision of a liberalism centered on public, rather than merely private, liberty.

Polanyi began to publish some work on economics in this period. For instance, he made a number of trips to the Soviet Union and took detailed notes on economic matters, which he combined with an analysis of the

[3] See Scott and Moleski (2005) Chap. 6, and Nye (2011) Chaps. 3, 4, 5, 6, 7.

[4] See Beddeleem, Martin (2019) "Michael Polanyi and Early Neoliberalism," *Tradition and Discovery* 45/3: 31–44.

nation's first two Five Year Plans in a lengthy critique of central planning.[5] There was a strong push toward economic education for the layperson in Polanyi's work of this period as well.[6] He also made his first published foray into philosophical matters with a paper in 1936 defending the semi-empirical method which he had helped to pioneer in chemistry.[7]

1.4 The Transition to Philosophy[8]

Polanyi stopped writing explicitly economic works by the end of the 1940s and began to focus more exclusively on philosophical topics. Nye points out that his "economic preoccupations form a bridge to his sociologically inflected philosophy of science" (2011, 176) since "he had found in economics an important way of thinking about the institutional framework for science" (Ibid., 178). Thus for Polanyi, science required institutions that were essentially liberal in character, in order to allow for the free flow of ideas and funding to create a spontaneous polycentricity that would allow for the pursuit of scientific truth.[9] In all this, Polanyi seems a standard liberal, however he also emphasizes public over private liberty and insists that the scientific "system is one of community," similar to his left-wing brother, Karl Polanyi's "description of preindustrial societies, where reciprocity, redistribution, and community were all important" (Ibid., 179). Thus,

> The key lay in focusing on functions, institutions, and social relationships as well as on individuals acting in isolation. This was a shared point of emphasis with Karl's *Great Transformation*, namely, the objection to

[5] Polanyi, Michael (1935). "U.S.S.R. Economics: Fundamental Data, System, and Spirit." *The Manchester School* 6/2

[6] See Polanyi (1937), "On Popular Education in Economics," made available in print in July 2016 in *Tradition and Discovery* 42/3: 18–24.

[7] Polanyi, Michael (1936). "The Value of the Inexact." *Philosophy of Science* 3: 233–234.

[8] See Scott and Moleski (2005) Chaps. 7–8 and Nye (2011) Chaps. 6, 7, 8.

[9] See Polanyi, Michael (1962b). "The Republic of Science: Its Political and Economic Theory." *Minerva* 1: 54–74.

positivist confidence that empiricism and reason alone can arrive through calculation at ideal economic arrangements." (Ibid., 178)

This rejection of positivism and pure empiricism along with sustained attention to the beliefs, values, methods, and phenomenological experiences of real scientists at work in the scientific community would ultimately drive Polanyi toward his development of a post-critical philosophy. The centerpiece of Polanyi's system of thought is a novel epistemology that emphasizes the personal and tacit dimensions of all knowledge as well as the related triadic attentional structure of all acts of knowing (knower, subsidiary details indwelt tacitly, and the comprehensive entity or meaning to which they point). This epistemology implies a metaphysics, which will be our primary concern.

In shifting our focus to the nuts and bolts of Polanyi's philosophy, it is important, I believe, to make clear the intensity of the existential and civilizational moral crises that provided the crucible within which his theorizing took shape. After all, it is striking indeed that a talented chemist with Nobel Prize-winning potential, a secure academic position, and a well-staffed and equipped laboratory under his sole supervision would allow his attention to drift to other matters. The explanation lies in the fact that Polanyi was motivated in the pursuit of this philosophical project by his keen awareness of the unstable moorings of Western society, having lost family members in the twentieth century's totalitarian experiments and having himself been twice-exiled from lands ravaged by revolutionary upheaval.

He made this change in intellectual focus because he believed that, to a significant extent, the atrocities that rocked Europe and the world in the first half of the twentieth century could be traced to the delusional idealization, both by the general public and many scientists themselves, of detached and impersonal scientific objectivity. In short this is because if such a detached objectivity is held to be the preeminent prerequisite of knowledge, then moral standards will not make the cut, and this will lead to widespread political violence.

For instance, Polanyi points out that if we consistently apply the standard of doubt argued for by Locke and others, which holds that it is irresponsible and perhaps reprehensible to impose beliefs that cannot be demonstrated, then.

> it follows that, unless ethical principles can be demonstrated with certainty, we should refrain from imposing them and should tolerate their total denial. But, of course, ethical principles cannot, in a strict sense, be demonstrated: you cannot prove the obligation to tell the truth, to uphold justice and mercy. It would follow therefore that a system of mendacity, lawlessness, and cruelty is to be accepted as an alternative to ethical principles and on equal terms. But a society in which unscrupulous propaganda, violence, and terror prevail offers no scope for tolerance. Here the inconsistency of a liberalism based on philosophic doubt becomes apparent: freedom of thought is destroyed by the extension of doubt to the field of traditional ideals, which includes the basis for freedom of thought. (*M* 9–10)

Likewise, moral standards might be critiqued on political grounds:

> The genuineness of moral standards [is] rendered suspect when it is realized that they are upheld by force, based on property and imbued with local loyalty...Morality will then be reduced to a mere ideology, and this depreciation of thought will tend to spread and to bring about eventually the subjection of all thought to local patriotism, economic interest and the power of the state. (*PK* 216)

This appraisal would then be taken as an objective, scientific critique of the validity of moral standards, yet paradoxically.

> These supposedly scientific assertions are...accepted only because they satisfy certain moral passions. We have here a self-confirmatory reverberation between the theory of bourgeois ideologies and the concealed motives which underlie it. This is the characteristic structure of what [Polanyi calls] a dynamo-objective coupling. Alleged scientific assertions, which are accepted as such because they satisfy moral passions, will excite these passions further, and thus lend increased convincing power to the scientific affirmations in question—and so on, indefinitely. (*PK* 230)

In Polanyi's analysis this scientistic philosophical background set the stage for the emergence of nihilism, first at the private, individual level, (as seen in the character of Raskolnikov in Dostoevsky's *Crime and Punishment*, to cite one of his favorite literary examples). Such individual nihilists "are experimenting privately with a life of total disbelief. But within a few years we see the nihilist transformed into…an ice-cold businesslike…political conspirator… closely prefiguring the ideal Bolshevik as…seen…represented on the Moscow stage in the didactic plays of the early Stalinist period" (*M* 15).

Polanyi paints a more granular picture of the rise of nihilism from the objectivism flowing out of the critical Enlightenment orientation by tracing the evolution of the unstable Raskolnikovian individual nihilist who is "liable to be drawn into politics, provided he can find a movement based on nihilistic assumptions" (*M* 16) through to the organized and orchestrated nihilism of the German Youth Movement and other currents within twentieth century politics that adopted "a creed of political violence" (Ibid.).

When faced with the prospect of supporting or directly carrying out acts of atrocious political violence, such individuals tacitly made use of the dynamo-objective coupling described above to justify themselves to themselves. For, while "their nihilism had prevented them from demanding justice in the name of justice or humanity in the name of humanity… their [repressed] moral aspirations…found an outlet in the scientific prediction of… a scientific utopia, relying for its fulfillment only on violence (Ibid., 17).

Thus,

> It was a mistake to regard the Nazi as an untaught savage. His bestiality was carefully nurtured by speculations closely reflecting Marxian influence. His contempt for humanitarian ideals had a century of philosophic schooling behind it. The Nazi disbelieved in public morality the way we disbelieve in witchcraft. It is not that he had never heard of it; he simply thought he had valid grounds for asserting that such a thing cannot exist. If you told him the contrary, he would think you peculiarly old-fashioned or simply dishonest…In such men the traditional forms for holding moral ideals had been shattered and their moral passions diverted into the only channels

> which a strictly mechanistic conception of man and society left open to them. We may describe this as a process of moral inversion. The morally inverted person has not merely performed a philosophic substitution of material purposes for moral aims; he is acting with the whole force of his homeless moral passions within a purely materialistic framework of purposes. (Ibid., 17–18)

The scope of the philosophical and social problems that preoccupied Polanyi are huge. By grappling with them for a few pages we begin to grasp the all-encompassing stakes involved in his philosophical work. Polanyi believed that a properly post-critical epistemology and the metaphysics (i.e. moral realism) it implied had the potential to renew Western Civilization, and it was to this project that he dedicated the rest of his intellectual life. His formal epistemological work began in the 1950s and continued into the 1970s, shortly before his death in 1976. Reception of his philosophical output was mixed, however his theory of tacit knowing prefigured the social turn in philosophy of science and has been influential in fields ranging from knowledge management to religious studies.[10]

Polanyi, then, is arguably best remembered at this point as an epistemologist of some note who operated outside the canons of academic philosophy. His principal influence outside of chemistry has been through his systematic theorization of the tacit dimension of knowing, yet when surveying his corpus, the epistemology he developed is never put forward as his intended destination or resting point. Rather, it is a necessary component of the movement from the critical milieu of modernity to a post-critical orientation within which metaphysical claims (including those

---

[10] Despite his relative philosophical success, Polanyi was disappointed that his contributions did not kickstart a noticeable rebirth of Western thought during his lifetime. Wigner and Hodgkin note that

> Though Polanyi was pleased by the reception which many of his ideas received in the United States, in [his later days in] Oxford there was sometimes a disappointed note in his conversation. ...Why was it that people were not taking more notice of his ideas? And, especially in Oxford, why were the traditional pillars of positivism and of linguistic philosophy not showing more signs of bending or yielding when he and his continental allies – Merleau-Ponty for example – had done them the honour of exposing some of their weaknesses? (1977, 434)

that undergird the practice of contemporary science) could once more be made in good faith, despite our lack of a certain, explicitly articulable foundation for such claims. It is thus to Polanyi's post-critical approach to philosophy that we now turn.

2 The Post-Critical Approach to Philosophy

2.1 The Critical and the Post-Critical

Polanyi coins the term post-critical[11] to describe a philosophical orientation that unapologetically takes the fiduciary nature of all knowing as its point of departure, thus rejecting modernity's critical approach which takes methodological doubt to be the proper epistemic fundament. For a post-critical thinker, then, methodological belief is foundational, and doubt, though it has its proper and necessary place in the pursuit of truth, plays second fiddle, fulfilling an important yet complimentary role. According to Polanyi, the fiduciary basis of all knowing is rooted in the inescapable personal and tacit coefficients of knowledge, which we will examine in more depth in Sect. 2.2, when we explore his epistemology in some detail.

The post-critical project was further developed by William Poteat, one of Polanyi's early philosophical admirers, who did his best to meet head on the challenge posed by Polanyi's magnum opus, *Personal Knowledge: Toward a Post-Critical Philosophy* in works such as *Polanyian Meditations: In Search of a Post-Critical Logic.* Poteat, perhaps more than Polanyi, noted post-critical themes and resonances in the works of other contemporary thinkers such as the late Wittgenstein, Hannah Arendt, and the phenomenological tradition, especially the work of Merleau-Ponty.

In the time since Polanyi and Poteat's heyday, Dale Cannon, himself a student of Poteat, has done some of the most comprehensive and penetrating work not only synthesizing and analyzing Polanyi and Poteat's

---

[11] It is important to note that other contemporary thinkers sometimes use the term "post-critical" to refer to positions (such as Kantianism, for instance) that in Polanyian terms are thoroughly critical. See, for example, Quentin Meillasoux (2006, 4).

conceptions of the post-critical, but also broadening and making more explicit the links to analogous and supporting work from thinkers working on the extended mind, feminist theories of knowledge, philosophical anthropology, and certain areas of sociology, seeing these as potential intellectual partners in the pursuit of a post-critical paradigm (Cannon 2018, 7). We will return to Cannon's incisive analysis shortly, but first we will direct our attention to several passages from Polanyi's *Personal Knowledge*[12] that are particularly helpful in making clear the distinction between the post-critical and the critical. Polanyi describes the critical turn in philosophy in the following terms:

> It has been taken for granted throughout the critical period of philosophy that the acceptance of unproven beliefs was the broad road to darkness, while truth was approached by the straight and narrow path of doubt. We were warned that a host of unproven beliefs were instilled in us from earliest childhood...We were urged to resist the pressure of this traditional indoctrination by pitting against it the principle of philosophic doubt. Descartes had declared that universal doubt should purge his mind of all opinions held merely on trust and open it to knowledge firmly grounded in reason. In its stricter formulations the principle of doubt forbids us altogether to indulge in any desire to believe and demands that we should keep our minds empty, rather than allow any but irrefutable beliefs to take possession of them...The method of doubt is a logical corollary of objectivism. It trusts that the uprooting of all voluntary components of belief will leave behind unassailed the residue that is completely determined by the objective evidence. Critical thought trusted this method unconditionally for avoiding error and establishing truth. (*PK* 269)

This trajectory leads to

> a steadily mounting distaste for certain forms of knowing and being; a growing reluctance to credit ourselves with the capacity for personal knowing, and a corresponding unwillingness to recognize the reality of the unspecifiable entities established by such knowing. (*PK* 350)

[12] Cannon (Ibid., 2–3) points to some of these passages as important delineators of the genesis of the post-critical project.

In contrast,

> The alternative…which I [Polanyi] am seeking to establish…is to restore to us once more the power for the deliberate holding of unproven beliefs. We should be able to profess now knowingly and openly those beliefs which could be tacitly taken for granted in the days before modern philosophic criticism reached its present incisiveness. (*PK* 268)

The aim, then, "is to re-equip men with the faculties which centuries of critical thought have taught them to distrust" (*PK* 381), and in order to do this, we might look toward historical precedents for such a move, by looking back toward the fourth century A.D. and St. Augustine, who

> brought the history of Greek philosophy to a close by inaugurating for the first time *a post-critical philosophy*. He taught that all knowledge was a gift of grace, for which we must strive under the guidance of antecedent belief: *nisi credideritis, non intelligitis*…We must now recognize belief once more as the source of all knowledge. Tacit assent and intellectual passions, the sharing of an idiom and of a cultural heritage, affiliation to a like-minded community: such are the impulses which shape our vision of the nature of things on which we rely for our mastery of things. *No intelligence, however critical or original, can operate outside such a fiduciary framework.* (*PK* 365f emphasis added)

The upshot, as Polanyi notes, is that,

> Any enquiry into our ultimate beliefs can be consistent only if it presupposes its own conclusions. *It must be intentionally circular.* The last statement is itself an instance of the kind of act which it licenses. For it stakes out the grounds of my discourse by relying essentially on the very grounds thus staked out; my confident admission of circularity being justified only by my conviction, that in so far as I express my utmost understanding of my intellectual responsibilities as my own personal belief, I may rest assured of having fulfilled the ultimate requirements of self-criticism; that indeed I am obliged to form such personal beliefs and can hold them in a responsible manner, even though I recognize that such a claim can have no other justification than such as it derives from being declared in the very terms which it endorses. (*PK* 299 emphasis added)

This fiduciary framework is foundational for a post-critical orientation, and its mere recognition ought logically to be the undoing of naïve scientism, since it undermines the objectivism presupposed by scientism[13] by demonstrating that "knowledge in all of the sciences, even the most formal and rigorous, is built in practice upon an acritical reliance upon ordinary embodied perceptual interaction with things in the world around us" (Cannon 2016, 35) and is thus an unmasking of the methodological faith at the heart of the scientific process (one thinks of Walter Benjamin's analogy of the dwarf in the Mechanical Turk as the theological heart of Marxism).[14] Put another way, the post-critical framework shows that the central, if often tacit, presupposition of the critical project – namely "that observation is transparent, that it is possible to perceive without also apperceiving" (Lewin 1991–1992, 19) is false, though it does this without succumbing to the sort of proto-nihilistic skepticism often operating in post-structuralist critique.

There are two crucially important points I wish to draw our attention to here. The first is that an open acknowledgement of the methodological faiths that underpin all of our pursuits leads to the anti-scientistic conclusion that "to be post-critical is to recognize that the natural sciences stand not above but alongside, and rightfully in complementary relationship with, the different kinds of inquiries pursued among the arts and the humanities, as well as in manual trades and commonsense experience" (Cannon 2016, 28). The second is the emphasis on the embodied nature

---

[13] In Polanyi's corpus scientism and objectivism are treated as different facets of the critical project which typically come intertwined, the one offering circular justification for the other and vice versa.

[14] In the first of his *Theses on the Philosophy of History* (1940), Benjamin writes:

> It is well-known that an automaton once existed, which was so constructed that it could counter any move of a chess-player with a counter-move, and thereby assure itself of victory in the match. A puppet in Turkish attire, water-pipe in mouth, sat before the chessboard, which rested on a broad table. Through a system of mirrors, the illusion was created that this table was transparent from all sides. In truth, a hunchbacked dwarf who was a master chess-player sat inside, controlling the hands of the puppet with strings. One can envision a corresponding object to this apparatus in philosophy. The puppet called "historical materialism" is always supposed to win. It can do this with no further ado against any opponent, so long as it employs the services of theology, which as everyone knows is small and ugly and must be kept out of sight.

of our knowing (what William Poteat calls our "linguistically shaped and embrangled mindbodily being in the world") (1985, 42).

Given that what Polanyi calls the critical turn in Western thought is defined by an unhealthy exaltation of the role of doubt in our process of coming to know and that his search for a post-critical philosophy begins with his reconstituted epistemology, we will now take a more detailed look at his contributions in this area before returning to the contours of the post-critical project writ large. *Nota bene:* In the arguments following this chapter, I shall operate under the assumption that Polanyi's approach to epistemology is generally correct and will freely use its terminology and trajectory in the development of my own project.

2.2 Polanyi's Post-Critical Epistemology

Polanyi proposes a realist epistemology centered around a triadic attentional structure which undergirds all acts of knowing. The first piece of the triad is the knower – a person – whose personal existence is the locus of the noetic act. The second element is the set of tacitly indwelt subsidiary particulars or clues that point toward the third component, the focally-known comprehensive entity or meaning. Since all theories of epistemology, not least those critical theories unfolding under the influence of Descartes and Kant, concern themselves with the explicit, focally-known component of this triad, Polanyi's novel contribution is to point out and insist on the inescapability of the personal and tacit coefficients in all acts of knowing (skills, propositions, etc.).[15] Taken together, the personal and tacit coefficients constitute the fiduciary nature of all knowledge central to a post-critical orientation.

For Polanyi, the first member of the triad, the person, is the culmination (so far) of life's evolutionary process. He "regard[s] living beings as instances of morphological types and of operational principles

[15] Polanyi, much like Gilbert Ryle, wishes "to upset the assumed type-difference between thinking and doing" and to "prove that…knowledge-how is a concept logically prior to the concept of knowledge-that" (Ryle 1945–1946, 2–4). Despite this common ground, however, Polanyi points out that "Ryle does not have the conception of subsidiary [i.e. tacit] awareness" (*PK* 372) and that this prevents Ryle's position from breaking out of the critical paradigm.

subordinated to a centre of individuality" (*PK* 383). A living thing on this view is thus essentially a center of skilled achievement. Again, Polanyi notes that "living beings can be known only in terms of success or failure. They comprise ascending levels of successful existing and behaving (Ibid., 381). For Polanyi, then, the superior noetic powers displayed by human persons when compared, say, with chimpanzees, represent a difference in degree, not in kind. The common thread is the personal coefficient of knowledge, which is associated with a life form's pursuit of skilled achievement and is present not only in human knowing. In its varying manifestations it is present everywhere something succeeds in living:

> ...the appearance of the human mind has been so far the ultimate stage in the awakening of the world; and all that has gone before, the strivings of myriad centres that have taken the risks of living and believing, seem to have all been pursuing, along rival lines, the aim now achieved by us up to this point. They are all akin to us. (Ibid., 404)

To successfully navigate the completion of some achievement, the personal center will have to integrate the other two thirds of Polanyi's triad. To get a sense of how this works, think of the way we "inhabit" a bicycle as we use it to ride to the store. We feel the road through the bicycle without explicit, focal awareness of the bike itself; we are aware of it only insofar as it bears on the Gestalt meaning of the comprehensive act, which is the total unity of riding down the road in order to get to the store. In the moment of riding to the store we are only tacitly aware of the bicycle (and the contours of the road) while we are focally aware of the goal of arriving at the store to buy such-and-such, the meaning to which our tacit awareness of the bicycle, the road, and myriad other subsidiary details points.

Again, the ubiquity of this second element in the triad, the tacitly-known subsidiary details, is the foundation of the second constitutive facet of the fiduciary nature of knowledge insisted upon by post-critical thinkers, the tacit coefficient (often referred to as the tacit dimension) as well as the general from/to structure which Polanyi argues characterizes all knowledge. While there is some debate among Polanyi scholars about the ontological status of some varieties of integrated focal meanings, the majority view, which I will myself argue for and expand upon in due course, is that – while there are different types of *gestalten*, which sometimes indicate different types and levels of reality – Polanyi is a full-blooded metaphysical realist with regard to such *gestalten*.[16]

At this juncture it will be helpful to look at one of Polanyi's own examples in order to better understand some key details in his epistemology including the from/to attentional structure of all knowing, the phenomenology of the process of indwelling, and some salient details about the relationship of subsidiary details to comprehensive entities, especially with regard to the tacit, unspecifiable nature of some such subsidiary details. Polanyi explains that

> A striking feature of knowing a skill is the presence of two different kinds of awareness…When I use a hammer to drive a nail, I attend to both, but quite differently. I watch the effects of my strokes on the nail as I wield the hammer. I do not feel that its handle has struck my palm but that its head has struck the nail. In another sense, of course, I am highly alert to the feelings in my palm and fingers holding the hammer. They guide my handling of it effectively, and the degree of attention that I give to the nail is given to these feelings to the same extent, but in a different way. The difference may be stated by saying that these feelings are not watched in themselves but that I watch something else by keeping aware of them. I know the feelings in the palm of my hand by relying on them for attending to the hammer hitting the nail. I may say that I have a subsidiary awareness of the feelings in my hand which is merged into my focal awareness of my driving the nail. (*M* 33, emphasis added)

As will already be apparent, Polanyi's epistemology makes significant use of the tradition of Gestalt psychology, noting, as he does, that every act of knowing is constituted by two levels: The comprehensive entity (the driving of the nail) of which we are focally aware, and the subsidiary details – clues, constituent parts, etc. – of which we are only tacitly aware and which we integrate into our focal awareness through a process of

[16] This topic will first be treated in some depth in Chap. 2 and will reappear intermittently throughout the rest of this project. Distinctions between different types of *gestalten* will be handled in Chaps. 4 and 5.

indwelling, a skilled integrative action. According to Polanyi, we cannot be both tacitly and focally aware of the same entity at any given moment, though we may be able to oscillate back and forth between the two with reference to the same entity. So as long as I am focally aware of the driving of the nail, I am only tacitly aware of the hammer itself, its pressure against my hand, the ways in which my muscles are controlling the hammer, etc. As soon as I shift my focal attention to the weight of the hammer, say, and its pressure against my hand, my direct, focal awareness of nail-driving dissolves into the tacit background, and this may cause me to miss the nail entirely or strike it a glancing blow that causes it to bend. We might easily imagine similar shifts in attention with regard to the example of riding a bicycle, briefly outlined above. The lesson to be learned here is that subsidiary particulars point beyond themselves to the comprehensive, integrated unity that is their meaning and that the process of integrating these elements is a personal act of tacit indwelling.

Polanyi calls this ensemble of tacitly indwelt subsidiary particulars the tacit dimension, and it is furthermore important to note that not all such subsidiary details are focally specifiable at all. This means that there are some subsidiary details that can never be focally known as comprehensive entities unto themselves by the knower in question. Consider, for instance, the neuronal firings and the tensing and relaxation of individual muscle fibers that are subsidiary elements of the act of hammering a nail, but which the hammerer herself cannot isolate as points of focal awareness. Even if she does pause from hammering to ponder theoretically the necessity of such physio-chemical components, she can never *concretely and focally specify her own neuronal firings and muscle twitchings as tacitly integrated into her own hammering at the self-same moment of hammering.*

To better understand the unspecifiability of the tacit co-efficient of knowledge, consider the following illustration taken from Polanyi's essay "Faith and Reason," in which a distinguished psychiatrist places before his students a patient suffering from "a mild fit of some kind," (239) asking them to determine what type of seizure the patient was suffering from. After some debate, the psychiatrist settles the matter, announcing to his students "You have seen a true epileptic seizure. I cannot tell you how to recognize it; you will learn this by more extensive experience" (Ibid.). Polanyi continues…

> *The psychiatrist knew how to recognize this disease, but he was not at all certain how he did this.* In other words, he recognized the disease by attending to its total appearance and did so by relying on a multitude of clues which he could not clearly specify. *Thus his knowledge of the disease differed altogether from his knowledge of these clues.* He recognized the disease by attending to it, while he was not attending to its symptoms in themselves, but only as clues. We may say that he was knowing the clues only by relying on them for attending to the pathological physiognomy to which they contributed. So if he could not tell what these clues were, while he could tell what the disease was, this was due to the fact that, *while we can always identify a thing we are attending to, and indeed our very attending identifies it, we cannot always identify the particulars on which we rely in our attending to the thing.* (Ibid. emphasis added)

In addition to the necessary tacit co-efficient of all acts of knowing, Polanyi's example highlights the personal co-efficient, here centered on the psychiatrist as a competent center of achievement in the act of facilitating increasing competency with regard to the achievement in question on the part of his students, themselves personal centers of achievement. Polanyi thus demonstrates that any pair of subsidiary clues with their focal meaning "is not linked together of its own accord[, but rather that] the relation of a subsidiary to a focus is formed by the act of a person who integrates one to the other. The from-to relation lasts only so long as a person, *the knower*, sustains this" (*M* 37 emphasis added).[17] As we will see in more depth later, the personal sustenance given to such from-to relations flows out of the passional nature of what Polanyi calls our commitment situation.

Our personal participation in acts of knowing is ubiquitous, but it is not uniform or exercised in precisely the same way across all facets of human experience. For instance, according to Polanyi,

> …the bearing of natural science on facts of experience is much more specific than that of mathematics, religion or the various arts. It is justifiable, therefore, to speak of the verification of science by experience in a sense

[17] As will be argued later, this citation describes the knower's epistemological integrative efforts and the metaphysical integration that sustains entities unto themselves through time.

> which would not apply to other articulate systems. The process by which other systems than science are tested and finally accepted may be called, by contrast, a process of validation. Our personal participation is in general greater in a validation than in a verification. The emotional coefficient of assertion is intensified as we pass from the sciences to the neighbouring domains of thought. But both verification and validation are everywhere an acknowledgment of a commitment: they claim the presence of something real and external to the speaker. As distinct from both of these, subjective experiences can only be said to be authentic, and authenticity does not involve a commitment in the sense in which both verification and validation do. *(PK* 202)

Thus our personal participation in acts of knowing occurring in various domains of experience can roughly be characterized as verification (for science), validation (arts, religion, etc.), and authenticity (for subjective experiences). Yet due to the tacit nature of our knowledge, verification and validation represent the two ends of a continuum, rather than a strict, separable dichotomy. The distinction between both of these and the authenticity of truly subjective experience is that the validation-verification continuum involves a metaphysical commitment to the independent ontological standing of some entity or entities involved in the experience.

2.3 Post-Critical Orientation Revisited

Having clarified Polanyi's conception of the triadic and fiduciary structure of all knowledge, we will return now to some of its broader implications for post-critical thought. The first order of business will be to avoid some misunderstandings which are both common and understandable. For both Polanyi and Poteat, post-critical thought is neither a return to pre-critical approaches, nor is it a naïve or fideistic embrace of an uncritical perspective, which simply refuses to subject beliefs to appropriate scrutiny. As McCoy points out, Polanyi does not "reject the importance of the critical method, he only deprives it of [its] 'Imperial Mood' … and the assumption that it describes Reality 'objectively' from an 'ontological peak' outside historical, social location" (McCoy 2002–2003, 11). Doubt is not rejected out of hand, then, as an important epistemological tool, as we will see in more detail shortly.

Doubt's proper role, however, is limited by the recognition that what Polanyi calls *a-critical* elements are constitutive of our knowledge. Such recognition is, according to Cannon, "one of the defining characteristics of post-critical thought" (2016, 40). The a-critical category includes not only the unspecifiable subsidiary details with which we are already familiar, but also all those

> aspects of human knowing...including knowledge claims in the most formal and strict of the sciences...[that] cannot be rendered fully explicit in principle. [The term a-critical] refers to what cannot be subjected to critical reflection because it cannot be made explicit without distorting, dessicating, or destroying it. For example, the very foundational assumptions of critical reflection must be relied on and held a-critically; to subject them to critical suspicion is to destroy them and disable critical reflection. (Ibid.)

The critical paradigm refuses to face up to this relatively straightforward insight out of fear of falling prey to subjective credulousness as well as on the basis of an Enlightenment faith in the possibility (and supposed inevitability) of the eventual explicit rendering of all aspects of reality, which will supposedly vindicate the objective basis of the skeptical procedure. Such Enlightenment faith itself, however, is properly characterized as an a-critical fundament in the critical knower's epistemological structure. The post-critical approach then, "involves an *existential transformation*, a coming to and 'owning up' to oneself (a transformed self) as one now believes one truly is—a shift from personal absence to personal presence—as a responsible person, a person responsible for whatever is being claimed or presented, giving it backing" (Ibid., 25). Conversely, refusal to own up to our true noetic situation is a sort of bad faith.[18]

The post-critical move is to embrace and acknowledge in good faith the fiduciary foundations of knowing that we have been exploring and

---

[18] This is because, as Cannon (2016, 26) notes, "This notion of a personal, existential transformation, of course, implies that living under—or identifying with—the critical paradigm is in fundamental respects to miss and fall short of owning up to oneself as responsible person. In an important sense it is to be personally absent".

thus to “own up” honestly to our acritical fundaments including the cultural and linguistic traditions we grow up in and are dependent upon. This does not mean, however, that Polanyi and those influenced by him are thereby propounding a simplistic fideism. To the contrary, we openly acknowledge that “such *acritical* trust is not infallible” (Cannon 2016 34). We hasten to point out, however, that it is not blind either, but is rather “correctible” in the light of unfolding discovery (Ibid.). Furthermore, methodological faith is “directed not to explicit candidates for belief (even ones we happen to come up with ourselves) but to our access to this dimension and our investigative forays into it— i.e., to truth as transcendental: as apprehensible (in part) yet also transcending our own best efforts to render it explicit” (Cannon 2008, 79).

This results in a methodological shift of one hundred and eighty degrees, in which methodological faith – that is, the recognition that all possible acts of knowing are constructed on a fiduciary foundation of a-critically accepted elements of our mindbodily situation – is given pride of place. Meanwhile, methodological doubt is not simply rejected out of hand, but is rather given its proper role, subordinate to the operation of methodological faith, thus placing “the passion for creative discovery of new aspects of reality at the heart of the enterprise of knowing—in place of critical suspicion—alongside our other intellectual passions (e.g., for beauty and for justice) and creative motivations (Cannon 2016, 27).

Critical suspicion demands explicitation: “Show your work! All of it!” But we cannot completely comply with this demand in good faith since so much of our work is tacit, a-critical, and passional in nature. Such is the nature of the heuristics of discovery, even in the sciences – typically thought of popularly as the most austere bastion of explicitly critical method – as Polanyi so aptly shows:

> Such conclusion[s] may appear less strange if we consider the phases through which the propositions of science are usually brought into existence. In the course of any single experimental inquiry the mutual stimulus between intuition and observation goes on all the time and takes on the most varied forms. Most of the time is spent in fruitless efforts, sustained by a fascination which will take beating after beating for months on end, and produce ever new outbursts of hope, each as fresh as the last so bitterly

> crushed the week or month before. Vague shapes of the surmised truth suddenly take on the sharp outlines of certainty, only to dissolve again in the light of second thoughts or of further experimental observations. Yet from time to time certain visions of the truth, having made their appearance, continue to gain strength both by further reflection and additional evidence. These are the claims which may be accepted as final by the investigator and for which he may assume public responsibility by communicating them in print. This is how scientific propositions normally come into existence. The certainty of such propositions can differ therefore only in degree from that of previous preliminary results, many of which had appeared final at first and only later turned out to have been only preliminary. Which is not to say that we must always remain in doubt, but only that our decision what to accept as finally established cannot be wholly derived from any explicit rules but must be taken in the light of our own personal judgement of the evidence. (*SFS* 15)

All this can be a rather bitter pill to swallow for many of us who have breathed the air of the critical project nearly all our lives. It seems to pull the rug out from under the possibility of objectivity and plunge us into a naively subjective maelstrom where anything goes. But this is only because the objective/subjective dichotomy was a simplification offering a false choice in the first place – a simplification which "induces us to imagine ourselves not in an active exploratory relationship to reality but as confronted with competing explicit candidates for belief, which we are given critically to doubt or uncritically to believe" (Cannon 2008, 79).

The post-critical move is to again draw attention to the personal coefficient of all knowledge – the logic of achievement that presupposes and requires an active center exploring reality – and then to show the ways that the personal coefficient disrupts the subjective/objective binary. Polanyi argues that

> We may distinguish between the personal in us, which actively enters into our commitments, and our subjective states, in which we merely endure our feelings. This distinction establishes the conception of the personal, which is neither subjective nor objective. In so far as the personal submits to requirements acknowledged by itself as independent of itself, it is not subjective; but in so far as it is an action guided by individual passions, it is

> not objective either. It transcends the disjunction between subjective and objective. (*PK* 300)

For Polanyi and other post-critical thinkers, then, the subjective and objective distinction is not a simple question of either/or. Rather, the objective itself has, in a sense, subjective roots which are mediated through the personal. And this latter is distinguishable from the subjective in part due to its "universal intent" (*TD* 78) manifested through the acceptance of standards of rational conduct that function as an external pole offsetting the internal pole of the merely subjective. This acceptance of the external pole, however, is everywhere and always a *personally* grounded commitment, resulting in standards that are, in a sense, self-set, though not willy nilly. It is in this sense, then, an implicit metaphysical recognition of an ontology that transcends the self. Thus there is

> a mutual correlation between the personal and the universal within the commitment situation. The scientist pursuing an enquiry ascribes impersonal status to his standards and his claims, because he regards them as impersonally established by science. But his submission to scientific standards for the appraisal and guidance of his efforts is the only sense in which these standards can be said to pre-exist, or even to exist at all, for him. No one can know universal intellectual standards except by acknowledging their jurisdiction over himself as part of the terms on which he holds himself responsible for the pursuit of his mental efforts. I can speak of facts, knowledge, proof, reality, etc., within my commitment situation, for it is constituted by my search for facts, knowledge, proof, reality, etc., as binding on me. These are proper designations for commitment targets which apply so long as I am committed to them; but they cannot be referred to non-committally. You cannot speak without self-contradiction of knowledge you do not believe, or of a reality which does not exist. I may deny validity to some particular knowledge, or some particular facts, but then to me these are only allegations of knowledge or of facts, and should be denoted as 'knowledge' and as 'facts', to which I am not committed. (*PK* 302–303)

Again, this commitment is ultimately metaphysical in nature, even if only tacitly:

> For the scientist, having relied throughout his inquiry on the presence of something real hidden out there, will justly rely on that external presence also for claiming the validity of the result that satisfies his quest. As he accepted throughout the discipline which the external pole of his endeavor imposed upon him, he expects that others—if similarly equipped—will also recognize the presence that guided him. By his own command, which bound him to the quest of reality, he will claim that his results are universally valid. Such is the universal intent of a scientific discovery. (*TD* 78)

Note that implicit throughout these examples from Polanyi's corpus is the assumption that the knower making personal noetic commitments with universal intent does not exist in the isolation of Descartes' *cogito*, but rather in the hurly burly of mindbodies in reality. As such, an individual knower is always to some extent enmeshed in a community of knowers operating in a given cultural and linguistic setting that must be a-critically accepted for any thought or knowledge to take shape at all.

Furthermore, other members of this community of knowers (both individually and collectively) are accessible "via empathetic indwelling, commonsensically alongside other persons" – the individual knower is therefore "knowable and known in her knowing" (Cannon 2008, 81–82). Thus, "not only is a disciplined empathetic exploration of other perspectives requisite to objectivity, but persons who dedicate themselves to doing so responsibly may be said thereby to transcend their own subjectivity" (Cannon 2008, 89). Cannon, building on the work of Polanyi and Poteat, goes on to conclude that

> Responsible personal judgment is both objective and personal at the same time. One could say, then, that all knowledge is a human construction, as post-modernists would typically put it, but not all constructions are subjective; some can and should be asserted, as Polanyi put it, with universal intent and may then rightly be judged objectively true—i.e., judged that they actually connect with and engage objective reality in its transcendence. (Ibid.)

Polanyi refers to this communitarian cultural, linguistic, and educational background of tradition, along with our individual inclinations and temperaments as our *calling*. Our calling provides us with a sort of map of reality which we must tacitly indwell as a set of subsidiary details in order to navigate the world at all. Polanyi believes that we can focally attend to particular aspects of our calling and subject them to critique, though only on the basis of acritical reliance on other aspects of the selfsame calling. Our ability to call into question particular subsidiary details of our calling (though never the framework in its entirety at any one time) is what allows us to avoid, in limited and finite ways, confusing the map provided by our calling with the territory it describes (to borrow a metaphor flowing through Cannon's work) by making stepwise moves towards better aligning our map with the realities it charts not only through empathy with other knowers, but also through the exercise of what Polanyi calls the creative imagination, a "…dynamic intuition" (*CI* 7) that provides us "anticipatory powers" (Ibid., 9) – in other words, a tacit foreknowledge (*TD* 22–23) by which we follow a "gradient of deepening coherence [that] tells us where to start and which way to turn…eventually bring[ing] us to the point where we may stop and claim a discovery" (*CI* 7). Polanyi continues by noting that "the resemblance to the way the painter or the poet are guided by the hidden coherence they are seeking seems clear" (*CI* 7).[19]

Stanley Scott develops this idea compellingly in his excellent essay on applying the post-critical perspective to literature, further developing Polanyi's insight:

> …discovery happens as a result of activating the…imagination and intuition—in three stages: (1) indwelling the particular materials of study or inquiry; (2) breaking out of an accepted framework of ideas and practices; and (3) the moment of discovery itself, when the scientist or artist experiences an integration of particulars (ideas, facts, data, words, lines and colors, musical notes, etc.) into an intelligible form, as in gestalt formation.

[19] Esther Meek concisely sums up the stages of Polanyi's heuristics of discovery thus: "We can generate a model of the chronological or at least logical development of a discovery on the basis of Polanyi's description: 1. Dynamic intuition formulates problem, setting task to imagination; strategic intuition assesses value of task. 2. Imagination ransacks for clues. 3. Dynamic intuition subsidiarily assesses clues in terms of tacit focus, directs search as it senses proximity to solution. 4. Imagination ransacks for clues (2 and 3 can be repeated alternately a number of times). 5. Creative intuition spontaneously integrates clues into anticipated focus, achieving discovery. 6. Confirmatory intuition proclaims validity of discovery; imagination points to inexhaustible future manifestations (still ransacking for clues!)." (2017, 53)

> The end product of discovery can then in the case of a scientific theory, be tested for its veracity and ability to generate further research; or in the case of an artistic achievement, appreciated for the beauty of its formal properties, and its ability to evoke discovery in the existential realm. (Scott 2019, 18)

Thus, when we turn our focal attention to particular aspects of our calling, subjecting them to a controlled methodological doubt guided by an underlying methodological faith, dwelling-in sets the fiduciary stage for the operation of the creative imagination which allows us, at times, to break out of a paradigm (or to redraw our map, in Cannon's terminology) that we now recognize as insufficient. The post-critical owning-up to our true fiduciary state is therefore not a move toward static, fatalistic acceptance of the social milieu into which we've been thrown, but a recognition and embrace of the dynamic push-and-pull between tradition and discovery that forms the paradoxical yet symbiotic and relationally subversive core of a true traditionalism, as Cornel West and Simon Critchley recognize (2012).

Polanyi's incorporation of our calling into the structure of his post-critical philosophy is a recognition of the social embeddedness of all knowing. This, along with many of the other elements of the post-critical stance we have already explored, puts his work into natural conversation with post-structuralism and post-modern thought, more generally, though it also indicates important differences between Polanyi's approach and the general current of post-modern thought.[20] Lewin (1991–1992), in an investigation of the limitations of the post-critical understanding of objectivity points out that a consistent concern of the best work in the post-modern tradition is to interrogate the ways in which the socially embedded process of knowledge-creation can play a part in "delegitimat[ing] various groups and perspectives…in terms of the complicity to participate in the regimes of democratic inquiry through which the production of knowledge for social regulation is encouraged" (Ibid., 24). He goes on to argue that

[20] Cannon, using David Ray Griffin's terminology, identifies Polanyi's approach as being compatible with "constructive post-modernism," in contrast to "deconstructive post-modernism" (2008, 69–70).

> ...objectivity, even [post-critically understood] as passionate appropriation in the context of personal knowledge, *is severely compromised by a social embeddedness which it can afford neither to discount nor acknowledge, and so to which it must remain systematically blind.* Whether we ultimately rest easy with Polanyi's account of personal knowledge or come to historicize it within the peculiar coerciveness of modernity I leave as an open question. (Ibid. emphasis added)

I disagree with Lewin that the post-critical position must remain "systematically blind" to its social embeddedness. In fact, this embeddedness is consistently acknowledged by Polanyi and those influenced by him, (see, for instance, *SFS* 83, or McCoy 2002–2003, 11, cited above) and in this sense I think Lewin's argument misses the mark, though I do think he raises an important question.

Put differently, the question is whether Polanyi ever adequately addresses head-on the magnitude of the injustice presupposed by the actual societies which evolved out of the classically liberal milieu in which he was embedded and which he fought to uphold, and whether he did adequate business with the complex set of issues surrounding how the callings of interrelated subjects who find themselves on different rungs of the hierarchical social ladder are to be adjudicated.

This, it seems to me, is a fair question to ask, and my own view is that Polanyi does not confront these issues with the attention they deserve. Put another way, he does not clearly address questions of justice within his conception of a society of explorers in a sufficiently granular way, nor does he carefully consider those socially embedded individuals who are not on the vanguard of such exploration. In this sense, then, I agree with Lewin insofar as I believe Polanyi's own use of his proposed post-critical remedy is not thorough enough in these social spheres, but I do not agree with Lewin that this particular failure on Polanyi's part is indicative of a constitutive failing baked in to the post-critical project.

What Polanyi's post-critical contributions do is to begin a conversation in which the "pre-critical passion and faith underlying critical rigor and the potential of pre-critical traditionalism as a springboard for discovery" can be disclosed, "thus providing fiduciary roots in tradition and community for post-modern creativity" (McCoy 2002–2003, 9). The disclosure of these fiduciary and personal roots, along with the acknowledgement of the finitude of our individual perspectives achieved through our callings, which are in large part established through our thorough enmeshment in communities, provides a potential springboard for social reflection and transformation beyond that explicitly envisaged by Polanyi (and which may indeed allow for fruitful connections with the very different social vision of his brother, Karl Polanyi).

For example, Cannon, further developing the post-critical conception of objectivity rooted in Polanyi's epistemology, accentuates more emphatically the centrality to our knowing of "different perspectives and different procedures of finding out (i.e., the indefinite plurality of multiple inquirers into the same thing)" (2016, 30) and concludes that

> a true and adequate notion of objectivity recognizes this plurality—indeed, *presupposes* it—and requires it to be taken into account by always seeking *recognition in common* among competent and independent inquirers of the matters in question, not despite but in virtue of the multiplicity of perspectives brought to bear upon them. (Ibid. emphasis original)

It seems to me that the further development of the post-critical position represented by Cannon indicates how the post-structuralist critique raised by Lewin above might be fairly dealt with, though a thorough development of this theme is not a central concern of mine here.

Lewin, however, is in other respects deeply sympathetic to the post-critical project and raises his objections in order to refine and develop the conversation in a constructive direction. He makes a keen and important observation when he notes that

> Our living is not only within a paradigm community that legitimates a particular way of seeing and doing, as post-empiricists would have it. It is also an indwelling within a life-world that sustains a deeply personal sense of how things are, a metaphysic which guides our empirical inquiry. (Lewin 1991–1992, 20)

Lewin here draws our attention to a number of important facets of a post-critical orientation, and a bit of a gloss is in order. First, the intense and all-encompassing existential import of our being in the world – our mindbodily embranglement, as Poteat might put it. We exist – but we will not start from the austere and artificially restricted axiomatic space of the Cartesian *cogito*. Abstract thought will certainly be taken into consideration in our post-critical investigations, but so will the entire panoply of phenomenological experience, encompassing the full range of experience and personal participation as analyzed under the headings of verification, validation, and authenticity. Furthermore, such phenomenological experience can fairly be used as evidence for metaphysical judgements, since, after all, such experience is ultimately the basis for *all evidential judgements full stop*.[21] The post-critical position is that there is significant correspondence between knowing and being, so long as we are sure to expand our notion of "knowing" to adequately own up to the tacit and personal fiduciary foundations. In this sense we concur with Heidegger, who notes that "we cannot be transposed into [...metaphysics] at all, because *insofar as we exist we are always already within it*" (1998, 96 emphasis added).

Thus for post-critical thinkers, beginning with Polanyi himself, an epistemology which properly accounts for its own fiduciary dimensions is itself, properly understood, a foray into metaphysics and ontology which avoids the abstract, abstruse, and critical distance of much of the Cartesian and analytic approaches to metaphysics (which presuppose personal absence), and engages in the pursuit with the existential

[21] As I have pointed out elsewhere:

> Polanyi's...approach[e] to ontology vis-à-vis ordinary experience is in important respects similar to the eidetic reduction in the phenomenological tradition (e.g., Descartes's famous consideration of wax).
>
> [...His] approach to developing his post-critical epistemology takes actual, embodied, acts of knowing and analyzes them eidetically in order to arrive at a more adequate description of the essence of knowledge. Indeed, this might be taken as an extension of his training as a scientist: he is working to carefully examine what reality has laid before him, whether it be a quartz crystal or our noetic structure. In this sense the principles and techniques of good science and good phenomenology are one and the same. Husserl points out that "eidetic seeing holds no more difficulties or 'mystical' secrets than does perception" (Husserl 2002, 272). (Turkis 2019, 33–34)

intensity demanded by personal presence implied by the fiduciary foundations.[22]

As we will see in subsequent chapters, not all post-critical interpreters of Polanyi's corpus are in agreement as to the contours of the metaphysics thus implied, and the bulk of my argument will be concerned with arguing that a version of Platonist metaphysics is either already presupposed by Polanyi's approach (though largely unbeknownst to him) or is a necessary complement to his epistemology if his proposed ontology is to get off the ground.

The details of the sort of Platonist metaphysics I am advocating will be forthcoming more completely later, however I think it appropriate to note here that my interpretation of Plato rejects any hard, Two Worlds division between the concrete and abstract, and I include here a sample of the sort of inclusive metaphysics found within this stream of Platonic thought in order to accentuate its coherence with the post-critical embrace of epistemology as beginning with phenomenology, and phenomenology (properly carried out *sans* Cartesian baggage) as metaphysics, that we have been discussing here.

> According to this…tradition every entity is a metaphysical entity: the toenails I cut this morning, the thank-you letters I should have written, the emotion of love, the thoughts that go through my head as I write this - everything. Plato's metaphysics is a prime example of such an inclusive metaphysics. The world[23] of Becoming may be held to be somehow less real than the world of the Forms, but the important point here is that the reality in which it participates is still the same metaphysical kind of reality that the Forms are thought to have. Objects in the world of becoming have their imperfect reality through participation in the Forms. (Robjant 2000)

[22] See Charles Lowney (2013–2014).

[23] I read "world" here as "realm" or "sphere" since those terms are more commonly understood as co-existing within the same world. I believe that something like this is Robjant's intention here, though he follows the use of "world" common in work on Plato. The connotations associated with "world," outside the sphere of Plato scholarship, however, often lead thinkers less familiar with his corpus to assume that the only possible interpretation of Plato is some form of a reified Two Worlds Theory, which is not the case, as will be argued more fully later. "World" is here analogical in the same way it is in statements like "There are a variety of subcultures to be found in the world of skateboarding," or "The world of political office is full of backbiting and intrigue."

Cannon notes that "to be post-critical is always first to exit the theater of [critical, Cartesian] solitude, to step outside of this uprooted posture of reflection, and recover one's convivial, commonsense mindbodily ground in order to be in a position to regard and assess these abstracted products of critical reflection for what they are in their true light" (2016, 34). To do so is to navigate the realities of toenail clippings, unwritten letters (and the potentially important social consequences thereof), moral conundrums, political systems, economic realities, and debates over universals and particulars. All of it is experienced and known through the existential immediacy of first-person, conscious, unified, mindbodily entanglement with the world and with others. Taken together, it (all of it) makes up the life-world we inhabit. And all of it is, in the final analysis, metaphysical on this inclusive view.

The metaphysical implications of our phenomenological experience are connected in important ways to a topic to which we will turn in more detail later, namely, Polanyi's definition of the real – that which has the power to continually surprise us through its *indefinite future manifestations* (IFM hereafter, following Esther Meek's usage) (2017, 77). I consider the IFM effect, as Meek calls it, to be a form of and interchangeable with the Eleatic Principle, that is to say, the requirement that an entity be causally potent in order to have its ontological status recognized, so-called due to its affirmation by the Eleatic Stranger in Plato's *Sophist* (247d–e). Causal potency, we should note, is taken here broadly so as to include formal, final, material, and efficient causation as legitimate ontological markers.

When this union of epistemology, phenomenology, and inclusive metaphysics properly understood, it lends an existential urgency to post-critical thought. In laying bare and owning up to the fiduciary foundations of all knowing, it makes central to our being trust and methodological faith and thus sets the stage for a clearing away of the critical and scientistic idols of absence and detachment, which caustically eat away at our sources of meaning. The alternative to embracing this type of live methodological faith, as Cornel West points out, is to allow "deception to be hegemonic" (Critchley and West 2012).

2.4 The Post-Critical Ethos, Plato, and the Aporia

Few of us, if any, can function under the assumption that deception is hegemonic. Thus the existential import of our direct experience ought naturally to drive us to recognize the fiduciary foundations of our phenomenological experience, already shot through with the ontological, leading us thereby to the realization that we are in the midst of a lived-out metaphysics, a way of being in relation with reality and other beings. Following this thread, Cannon notes that

> Polanyi's words imply that he was seeking to articulate a post-critical philosophy… However, it seems clear that it was not just toward a post-critical *philosophy* that Polanyi was aiming. Just as much or even more so, I believe, Polanyi was seeking to articulate a vision…[and] foster the emergence…of a *post-critical intellectual ethos*, a context and style of intellectual life, a "convivial order," that would be free of the inordinate critical passions and objectivist epistemology that plague the modern critical ethos and render it so problematic and unconvivial…This shift…is meant to broaden the focus from the individual knower in the abstract to the knower in community with other knowers… (Cannon 1992–1993, 5 emphasis original)

Poteat often points out that we both have and are in the midst of a picture (1985, 15 and throughout). I have been attempting to paint a picture of what the post-critical posture or ethos is, but I also want to declare that it is the picture that I am (striving to be) in the midst of. Thus it is from within this picture and with such a post-critical ethos that I will, in subsequent chapters, examine Polanyi's more explicit ontological claims as well as some of his more evocative and cryptic metaphysical formulations. The question of Polanyi's views on universals and particulars will be of particular interest. My purpose will be to show that Polanyi's epistemological insights, which I take as givens, are pre-figured by and harmonious with a way of understanding Platonism, and that something like this Platonist stance is a necessary background if Polanyi's post-critical metaphysical project, which he approaches through his epistemology, is to get off the ground. I will therefore be advocating for a post-critical Platonism.

While the details of the post-critical orientation are freshly before us and before turning our attention to the specifics of Polanyi's explicit metaphysics (as opposed to the implicit Platonist metaphysic that remains tacit in his work and that I aim to disclose), I think it wise to mention several ways in which Plato's work fits well with the post-critical ethos – a better fit, for instance, than certain forms of Aristotelianism that often tacitly inform Polanyi scholarship.

The first of these is the natural aporetic openness of the post-critical ethos, which clearly acknowledges any number of aporia. Yet the post-critical thinker, following up on Polanyi's insights, does not attempt to bypass or to resolve such paradoxes and apertures, but rather to dwell in them, thereby exploring and elucidating them in sane and healthy ways, and asking thereby whether there is "...any...way forward in which our deepest convictions might be whole-heartedly affirmed with intellectual integrity—fully realizing that we are finite and fallible? Is there hope for an integral reformation of our intellectual conscience?" (Cannon 2008, 77). Such a way forward is taken to present itself in the form of the existential owning up we have been discussing here for some time.

Polanyi's aporetic openness is on full display when, in discussing the limitations of our socially-embedded callings, he states that he accepts it as

> ...inevitable that each of us must start his intellectual development by accepting uncritically a large number of traditional premisses of a particular kind; and that, however far we may advance thence by our own efforts, our progress will always remain restricted to a limited set of conclusions which is accessible from our original premisses" (*SFS* 83)

Furthermore, there is the inevitable aporia that opens up when we acknowledge the tacit dimension of our knowing and own up to the fact that many of the subsidiary particulars upon which we depend in our from-to (sometimes referred to as "from-via-to" or "from-through-to" relationships) [24] are unspecifiable, yet indispensable.

[24] On "from-via-to" see Gulick, "The Meaningful and the Real in Polanyian Perspective (1999, III). On the "from-through-to" structure and the transparency and indeterminacy of the "through" see Lowney, "The Tacit in Frege" (2008, 28) "Epistemontology" (2013–2014, 21), or Lowney (2017, 159–160).

Another side of the aporetic resonance of the tacit dimension is gestured toward by Poteat when he speaks of our tacit ability to "seamlessly integrate" an expert opinion – say, the explicitly delivered and competent analysis of some maladroit physical motion we have been carrying out –

> …to our own motor acts without even vague maxims for how it is we are to accomplish this integration. And even saying how it was done after the fact would be, under the best of circumstances, no small feat. (Poteat 1985, 168)

Polanyi also openly recognizes and embraces the aporia inherent in the epistemological circularity of all axiomatized systems, ranging from mathematics to the native cultural-linguistic system into which we are thrown as participants upon our births (*PK* 288–290). Yet such recognition does not lead him to jettison the validity of mathematics or of all conclusions reached through our personal indwelling of such systems. This tension between the limited and finite nature of our calling and the demand from within that calling to responsibly hold our beliefs with universal intent is rather formulated in the naturally aporetic terms, already discussed, of his theory of personal knowledge.

Furthermore, Polanyi's theorizing on the heuristics of discovery begins, perhaps, with his analysis of Plato's *Meno (1997a)*, where he notes:

> …we still have no clear conception of how discovery comes about. The main difficulty has been pointed out by Plato in the Meno. He says that to search for the solution of a problem is an absurdity. For either you know what you are looking for, and then there is no problem; or you do not know what you are looking for, and then you are not looking for anything and cannot expect to find anything. If science is the understanding of interesting shapes in nature, how does this understanding come about? How can we tell what things not yet understood are capable of being understood? The answer I [give] here to this question [is] that we must have a foreknowledge sufficient to guide our conjecture with reasonable probability in choosing a good problem and in choosing hunches that might solve the problem. (*SFS* 6)

Thus Polanyi does not balk at the aporetic logical gap which separates the investigator from the discovery and which is crossed by means of heuristic passion following tacit foreknowledge of novel integrations (the moment of breaking out and of re-integration briefly discussed earlier), despite the fact that the tacit roots of such discovery make an exhaustive explanation of the process impossible, even in retrospect. Likewise, the discoverer, having crossed Meno's Gap finds herself within an altered calling, for no longer is she in agreement with the received view, having broken out of the old paradigm. Yet another aporetic gap here presents itself, for "to the extent to which [she] has committed [her]self to a new vision of reality, [she] has separated [her]self from others who still think on the old lines" (*PK* 150). Thus, her "persuasive passion spurs [her] now to cross this gap by converting everybody to [her] way of seeing things" (Ibid.). This is the "aporia revealed at the point where an individual emerges from his formative traditions" (Lewin 1991–1992, 23).

One aspect of the post-critical embrace of such aporia is that recognition of the fiduciary foundations of all human knowing leads to an existential embrace of metaphor. This is because, when "knowing" is given its existential due and taken to include all skilled integrative activity, to recognize the methodological faith at the heart of knowing is to recognize the fiduciary foundations of *all distinctively human activity* – activity which is, at a constitutive level, the mobilization of various types of "metaphors, metonymies, synecdoches," as Cornel West points out (Critchley and West 2012).[25] Polanyi, though he regularly compares scientists to writers and artists, does not, perhaps ever completely follow this thread to the end, for though he works on a theory of metaphors in *Meaning*, noting that "a metaphor, like a symbol, carries us away, embodies us in itself, and moves us deeply as we surrender ourselves to it," (*M* 78) he also problematically raises some "inchoate ideas [about…] rigid distinctions between art and non-art, indicative and symbolic, 'open-eyed' and political" (Bocharova 2019, 10). Nonetheless, on balance, Polanyi affirms that

[25] If West's claim seems a bridge too far, consider the insights of the burgeoning field of biosemiotics, which studies the most basic biological processes as instances of sign-deciphering, right down to the enactment of the information contained in DNA.

> Dwelling in aesthetic visions is deeply persuasive, influencing what we are able to judge as plausible. The experience of the transcendent—achieved through transnatural[26]integrations—may potentially re-order our deepest commitments… (Ibid., 10)

Polanyi, then, provides a post-critical impetus toward recognizing metaphor in its various forms as a heuristic guide to reality, meaning that art and literature are "truth-seeking and truth-revealing activities," as Iris Murdoch (1997 location 575) might put it.

Poteat, perhaps due to his assimilation of Wittgenstein, goes further here than does Polanyi, calling our attention to the "metaphorical intentionalities" that shape "the particularity of the bent, the tendency, disposition, intentionality, and 'logic' of a given picture" – a picture that we simultaneously have and are in the midst of (1985, 151). For Poteat, these metaphors are "tonic…they are the [very] vectors of the intentionality and tone of our mindbodily being" (Ibid., 155).

Stan Scott, in working out how our indwelling of metaphor can lead to the breaking out characteristic of discovery, notes that

> As a voice from outside logical space, literary metaphor makes little appeal to reason as commonly conceived, but speaks in a way that can break through the boundaries of existing logical systems. If read as the depth of speech, and not just dispensable ornamentation, metaphor can redefine what is meant by logic, by providing sets of premises that persuade us of unsuspected coherences in the existential realm. But it does this by diverging from existing logical criteria. (2019, 26–27)

Of course such metaphors can be for good or for ill. Cannon points out that "they can captivate our imagination and implicate us in abetting skeptical, relativist, reductionist, and even nihilist views without our realizing it" (2016, 33). On guard. The mobilization of methodological doubt with regard to metaphors is thus often appropriate from within the post-critical ethos, but again, this constrained skepticism is only possible within a larger backdrop of belief, shot through with metaphorical intentionalities.

---

[26] "…the imaginary entities created by means of the integration of incompatibles in art and myth…must be called *trans*natural" (*M* 125).

Such metaphors are in an important sense all we have. They are the tacit stuff of life and of death. We cannot escape pictures, but we must be careful with and about them, for "art is not a diversion or a side issue" (Murdoch 1997, location 6432). To own up in this wise is to take an important step toward achieving the balance of mind Polanyi sought.[27] This is the post-critical orientation from within which I will argue, and which I will not, generally speaking, argue for.

Plato, when read seriously, is of course deeply enmeshed in the aporetic paradoxes of our existence. First and foremost we have his nearly exclusive use of the dialogue form – a work of literary art as well as a philosophical treatise – even as he expels the poets from the ideal city. But of course this sort of aporetic contradiction is common in Plato's corpus. Plato himself, as himself, never quite appears in the works to clearly delineate his own views, even though his views, whatever they were, are key to kickstarting the Western philosophical tradition.[28] There are, of course, the early aporetic dialogues, but the aporetic interrogation extends throughout his *ouvre*. We might mention, to name just a few instances, the paradox at the heart of the *Phaedo* (to be discussed at length in due course) whereby our anamnetic access to the Forms is always activated through our contact with participating particulars (resonant also in the Socratic phenomenology which takes concrete cases as paradigm examples in order to get to the essence of a class or category), the consistent use of myth in later works, the seemingly contradictory or paradoxical conclusions reached across different dialogues, the withering critique in *Parmenides* of Plato's own theory of Forms which raises questions about *which* theory of Forms is the one to which Plato subscribed, etc.

Polanyi's acceptance of aporia and his emphasis on metaphor, when taken alongside the ambiguity in the theory of Forms in Plato's own corpus is indicative of a certain harmony between Plato (or at least certain interpretations of Plato) and the post critical project explicitly inaugurated by Polanyi which has yet to be fully fleshed out. While Colin Cordner (2017) thoroughly investigates the relationship of Polanyi's

---

[27] See Jon Fennell (2018).

[28] There is an evocative similarity here to the way in which the two volumes of Cervantes' *Don Quijote de la Mancha*, constituting the first modern novel, are also clearly the first post-modern novel due to their thoroughly self-referential intertextuality.

indwelling to Plato's *mimesis,* McCoy mentions only briefly "Plato's suggestions of a tacit dimension in human thought in the *Meno* and in the dialogical method that makes clear the human location of philosophical exploration" (2002–2003, 10). Poteat, though he rejects what he takes to be the Greek paradigm of thought as inferior to what he considers the Hebrew picture, in recognizing that Polanyi explicitly modeled his project on Augustine, draws attention to the fact that Augustine's thought was bathed in Neo-Platonism (1985, 140).[29] I thus take the possibility of a post-critical Platonism to be worthy of investigation.[30]

Furthermore, along with the greater prominence in analytic philosophy of Neo-Aristotelianism and contemporary hylomorphism, there has in recent years been a resurgence in interest in forms of Platonism (See McCabe 1994; Gerson 2004, 2005; Grant 2006; Berman 2020; Marmodoro 2021, etc.) or in theories that resemble Platonism to some degree, such as ontic structural realism. There is, however an important facet of classical Platonism that tends to be given short shrift by many of these analytic approaches, but which is properly at the center of the post-critical ethos, and that is the emphasis on Eros, that is to say, on beauty as a propulsive force.

Polanyi's heuristics of discovery, developed in the phenomenological light of his own experience in physical chemistry and his close proximity to other great scientific minds, begin with "a flow of passion" in which "*we recognize intellectual beauty as a guide to discovery and as a mark of truth*" (*PK* 300, emphasis added). Discovery begins with a passionate

[29] And thus, in my view, undermines the clarity of the distinction between the Greek and Hebrew pictures that he is relying on.

[30] Besides the Cordner (2017), Poteat (1985), and McCoy (2002–2003), mentioned above, additional scattered references to Polanyi's relationship to Platonism are extant, but there is no settled view on whether he is a Platonist. Edward Pols notes some alignment between Polanyi and Plato in "The Problem of Metaphysical Knowledge" (1968), but does not argue that he is a Platonist. Lowney (2008) compares Polanyi's view on universals to that of Frege, and notes that Polanyi "agrees with the rationalistic, or platonic…strain in Frege" (22). Lowney elsewhere refers to Polanyi more directly as a Platonist (2011–2012, 33) and includes a note on "what might be called…Polanyi's Neo-Platonism since…universals are real in [his] sense of allowing for future, and unexpected, manifestations" (2017, 166). Paul Richard Blum (2010) also makes several passing references to Polanyi as a Platonist. Fennell (2013), however, holds that Polanyi's understanding of universals does *not* posit "anything like a Platonic form of 'X,'" (373), and Gelwick (2013–2014) flatly states that "Polanyi is not a neo-Platonist" (28). My own (Turkis 2019) article on Polanyi and Iris Murdoch begins to make a more sustained argument for Polanyi's Platonism.

move towards some intriguing aspect of reality that we have yet to grasp. The passionate nature of this primordial move

> ...is unmistakably revealed by the emotional climax that accompanies discovery. Passions seek satisfaction and intellectual passions seek intellectual joys. The most general term for the source of this joy is beauty. The mind is attracted by beautiful problems, promising beautiful solutions; it is fascinated by the clues to a beautiful discovery and pursues untiringly the prospects of a beautiful invention. (Ibid., 36)

This "power of intellectual beauty to reveal truth" is a continuous theme in Polanyi's work (Ibid., 149), as Francisco Gallardo has argued (2003, 291–292). Polanyi notes, for instance, that an appreciation of abstract mathematical beauty leads to "the joy of grasping mathematics [and] induces the mind to expand into an ever deeper understanding... liv[ing] henceforth in active preoccupation with its problems" (*PK* 320). Meanwhile, within more observational sciences,

> ...passionate intimations of...harmonious order are the guides of discovery...and the beauty of a physical theory is the mark of its scientific value. This beauty is enjoyed by dwelling in the theory and observing its confirmation by the facts; the physicist dwells with pleasure in the patterns of inanimate nature, while he turns away coldly from disorderly, meaningless collocations of particles. (*SM* 74)

Such "scientific passion" seeking intellectual beauty "serves also as a guide in the assessment of what is of higher and what of lesser interest" and thus provides not only the primal impulse towards particular scientific disciplines, but also drives the adjudication amongst theoretical alternatives within a given science (*PK* 135).

Thus "the affirmation of a great scientific theory is in part an cxprcssion of delight. The theory has an inarticulate component acclaiming its beauty, and this is essential to the belief that the theory is true;" yet – again – this means that the structure of science is not so different from other human domains as modern orthodoxy would have us think since "a scientific theory which calls attention to its own beauty, and partly relies on it for claiming to represent empirical reality, is akin to a work of art which calls attention to its own beauty as a token of artistic reality" as well as to "the mystical contemplation of nature" (Ibid., 133).

As Lewin points out, what Polanyi recognizes as the passion that drives us to seek beauty "is akin to the erotic of Plato; it is the love for truth that motivates…To be passionate in inquiry is to prize the truth; it is to manifest a circumspection and care that, in its turn, is rewarded by that which is unveiled" (1991–1992, 21).[31]

A post-critical Platonism, then, will re-emphasize the existential power of Eros, the passionate personal commitment that is typically passed over by most contemporary Platonisms, but is fully acknowledged and theorized by Polanyi and his followers. It will also emphasize, deal with, and embrace the aporia, recognizing that such acknowledgement does not mean the end of learning, exploration, and engagement, nor the abandonment of the fruitfulness of science, etc., but is in fact part and parcel of the fiduciary foundations thereof.

For the fact that the critical milieu has disdained and ignored the tacit dimension of knowledge does not commit us as post-critical thinkers to giving the explicit dimension of knowledge – including metaphysical knowledge – short shrift. In the interest, then, of setting the stage for a constructive post-critical Platonist proposal, we will now dive into an exploration and critique of Polanyi's ontology and metaphysics, endeavoring to cover both those elements he theorized explicitly and those which are implied and hinted at by the metaphorical intentionalities of his texts.

3 Chapter Conclusion

After a brief examination of the scope of Polanyi's life and work we have attempted in this chapter to provide an understanding of Polanyi's post-critical ethos, with particular reference to developments of the project

[31] *Phaedrus* 250 d-e: "Vision, of course, is the sharpest of our bodily senses, although it does not see wisdom. It would awaken a terribly powerful love if an image of wisdom came through our sight as clearly as beauty does, and the same goes for the other objects of inspired [e] love. But now beauty alone has this privilege, to be the most clearly visible and the most loved." (Plato 1997c)

found in the work of William Poteat and Dale Cannon. A further function of this chapter is to serve as a sort of manifesto given that the rest of my project will unfold from within the ethos described here.

The core of the post-critical orientation is an explicit rehabilitation of methodological belief and a concomitant dethroning of the methodological doubt insisted upon by the critical turn in Western thought, though this does not imply any sort of fideism nor rule out the constructive role played by doubt when appropriately applied in our epistemic processes. This affirmation of the centrality of methodological belief has, as we have seen, existential and metaphysical consequences which find their nexus in our phenomenological experience as selves in the world.

Also included here is a brief explanation and exemplification of Polanyi's theory of tacit knowing, the systematized launchpad for his post-critical project that clearly outlines the fiduciary nature of all knowing. We have further emphasized that for Polanyi, the nature of our epistemic commitment is ultimately metaphysical since the continuous range of commitment – from verification to validation to authenticity – is primarily defined by the ontological weight given to the entities involved in a given phenomenological experience through the fiduciary participation of the knower.

Finally, some preliminary arguments in favor of the compatibility of Platonism with the post-critical ethos were laid out, specifically with regard to their shared emphasis on Eros and their openness to the inevitably aporetic nature of our existence.

Having thus prepared the way we will now turn our attention to Michael Polanyi's metaphysics, beginning with his Copernican realism.

References

Beddeleem, Martin. 2019 "Michael Polanyi and Early Neoliberalism," *Tradition and Discovery* 45/3: 31–44.

Benjamin, Walter. 1969. "Theses on the Philosophy of History." *Illuminations* trans. Harry Zohn. New York: Schocken Books: 253–264.

Berman, Scott. 2020. *Platonism and the Objects of Science.* New York, NY: Bloomsbury.

Blum, Paul Richard. 2010. "Michael Polanyi: the anthropology of intellectual history." Studies in East European Thought 62/2: 197–216.

Bocharova, Jean. 2019. "Toward a Post-Critical Literary Theory." *Tradition and Discovery* 45/1: 4–15.

Cannon, Dale. 1992–1993. "Toward the Recovery of Common Sense in a Post-critical Intellectual Ethos." *Tradition and Discovery* 19/1: 5–15.

———. 2008. "Beyond Post-Modernism via Polanyi's Post-Critical Philosophy." *The Political Science Reviewer* 37: 68–95.

———. 2016. "Being Post-Critical." *Recovering the Personal: The Philosophical Anthropology of William H. Poteat*, ed. by Dale W. Cannon and Ronald L. Hall. Washington D.C.: Lexington Books: 21–46.

———. 2018. "The Post-Critical Turn (How and Why Post-Critical Inquiry and Criticism Is Different)." Unpublished Presentation given at the Polanyi Society's Conference on the Post-Critical Turn. Nashotah, WI: http://www.polanyisociety.org/2018Nashotah/2018-Nashotah-Papers/Cannon-2018-NashotahPpr-6-4-18.pdf

Cordner, Colin. 2017. "Trust, Understanding, and Paradigms in the Work of Michael Polanyi and of Plato." *Appraisal* 11/3: 26–40.

Critchley, Simon and West, Cornel. 2012. "Faith of the Faithless: Simon Critchley and Cornel West in Conversation at BAM." https://www.youtube.com/watch?v=3oB2eQmVFLY

Fennell, Jon. 2013. "Polanyi, Universals, and the Nominalism Controversy." *The Journal of Speculative Philosophy* 27/4: 365–387.

———. 2018. "Balance of Mind: Polanyi's Response to the Second Apple and the Modern Predicament." *Tradition and Discovery* 44/2: 47–63.

Gallardo de la Torre, Francisco. 2003. "La Epistemología de Michael Polanyi: Una perspectiva realista de la ciencia." PhD diss., Pontificia Universitas Sanctae Crucis.

Gelwick, Richard. 2013–2014. "A Clue Toward Knowing Truth and God, and Polanyi's "Forms of Atheism."" *Tradition and Discovery* 40/2: 26–35.

Gerson, Lloyd P. 2004. "What is Platonism?" *Journal of the History of Philosophy.* Vol. 43: 253–276.

———. 2005. *Aristotle and Other Platonists.* Ithaca, NY: Cornell University Press.

Grant, Iain Hamilton. 2006. *Philosophies of Nature after Schelling.* New York, NY: Continuum.

Gulick, Walter B. 1999. "The Meaningful *and* the Real in Polanyian Perspective." *Polanyiana.* 8/1–2: http://www.kfki.hu/chemonet/polanyi/; http://www.ch.bme.hu/chemonet/polanyi/

Husserl, Edmund. 2002. "Philosophy as a Rigorous Science." *The New Yearbook for Phenomenology and Phenomenological Philosophy II.* Translated by Marcus Brainard. Taylor and Francis.
Lewin, Philip. 1991–1992. "The Problem of Objectivity in Post-Critical Philosophy." *Tradition and Discovery* 18/1: 18–26.
Lowney II, Charles W. 2008. "The Tacit in Frege." *Polanyiana* 1–2/17: 19–37.
———. 2011–2012. "Ineffable, Tacit, Explicable and Explicit: Qualifying Knowledge in the Age of 'Intelligent' Machines." *Tradition and Discovery* 38/1: 18–37.
———. 2013–2014. "From Epistemology to Ontology to Epistemontology." *Tradition and Discovery* 26/1: 16–29.
———. 2017. "Overcoming the Scientistic Imaginary." *Charles Taylor, Michael Polanyi, and the Critique of Modernity* ed. Charles W. Lowney II. Palgrave Macmillan. 143–168.
Marmodoro, Anna. 2021. *Form and Structure in Plato's Metaphysics.* New York, NY: Oxford University Press.
McCabe, Mary Margaret. 1994. *Plato's Individuals.* Princeton: Princeton University Press.
McCoy, Charles S. 2002–2003. "Ethics For The Post-Critical Era: Perspectives from the Thought of Michael Polanyi." *Tradition and Discovery* 29/1: 6–21
Meek, Esther Lightcap. 2017. *Contact with Reality: Michael Polanyi's Realism and Why It Matters.* Eugene, OR: Cascade Books.
Meillasoux, Quentin. 2006. *After Finitude: An Essay on the Necessity of Contingency.* Trans. Ray Brassier. New York: Bloomsbury.
Murdoch, Iris. 1997. *Existentialists and Mystics: Writings on Philosophy and Literature.* New York, NY. Penguin. Kindle Edition.
Nye, Mary Jo. 2011. *Michael Polanyi and His Generation: Origins of the Social Construction of Science.* Chicago: University of Chicago Press.
Plato. 1997a. *Meno* trans. G.M.A. Grube, in *Plato: Complete Works.* Edited by John M. Cooper. Indianapolis, IN: Hackett Publishing Company.
———. 1997c. *Phaedrus* trans. Alexander Nehamas and Paul Woodruff.
Polanyi, Michael. 1935. "U.S.S.R. Economics: Fundamental Data, System, and Spirit." *The Manchester School* 6/2.
———. 1936. "The Value of the Inexact." *Philosophy of Science* 3: 233–234.
———. 1961. "Faith and Reason." *Journal of Religion* 41/4 (Oct): 237–247.
———. 1962a [1958]. *Personal Knowledge: Towards a Post-Critical Philosophy.* Chicago, IL: University of Chicago Press.

———. 1962b. "The Republic of Science: Its Political and Economic Theory." *Minerva* 1: 54–74.

———. 1964 [1946]. *Science, Faith, and Society*. London: Oxford University Press.

———. 1966. *The Tacit Dimension*. Chicago, IL: University of Chicago Press.

———. 2016 [1937]. "On Popular Education in Economics." *Tradition and Discovery* 42/3: 18–24.

Polanyi, Michael and Prosch, Harry. 1975. *Meaning*. Chicago, IL: University of Chicago Press.

Pols, Edward. 1968. "The Problem of Metaphysical Knowledge." *Intellect and Hope*. ed. Thomas A. Langford and William H. Poteat. Durham, NC: Duke University Press: 58–90.

Poteat, William. 1985. *Polanyian Meditations. In Search of a Post-Critical Logic*. Durham, NC: Duke University Press.

Robjant, David. 2000. "Iris Murdoch's Everyday Metaphysical Entities." *Minerva – An Internet Journal of Philosophy* Volume 4, November 2000. http://www.minerva.mic.ul.ie/vol4/everyday.html

Ryle, Gilbert. 1945–1946. "Knowing How and Knowing That: The Presidential Address." *Proceedings of the Aristotelian Society*. New Series, Vol. 46: 1–16.

Scott, Stanley. 2019. "Indwelling and Breaking Out: Language and Literature in Post-critical Perspective." *Tradition and Discovery* 45/1: 16–29.

Scott, William T. and Moleski, Martin X. 2005. *Michael Polanyi: Scientist and Philosopher*. Oxford: Oxford University Press.

Turkis, Martin. 2019. "Post-Critical Platonism: Preliminary Meditations on Ethics and Aesthetics in Iris Murdoch and Michael Polanyi." *Tradition and Discovery* 45/1: 30–41.

Wigner, E. P. and Hodgkin, R. A. 1977. "Michael Polanyi. 12 March 1891--22 February 1976" *Biographical Memoirs of Fellows of the Royal Society* Vol. 23 (November 1977): 413–448.

Polanyi's Copernican Realism: Content, Reception, and Relation to Three Contemporary Realisms

1 Polanyi's Commitment to Metaphysical Realism

Polanyi is a realist, though precisely what sort of realist he is has given rise to some debate. In an issue of *Tradition and Discovery* dedicated to the topic of Polanyi's realism, Andy Sanders introduces the topic by noting that all of the issue's contributors

> ...agree that Polanyi advocated a scientific realism hanging on the theses that reality is independent of human conceptualizations and that it is partially and fallibly knowable...
>
> As to Polanyi's realism, it is undisputed that he advocates the independence thesis: reality "is largely hidden to us, and existing therefore independently of our knowing it" (*PK* 311). It is also undisputed that he maintains that human beings are able to come to know and understand the world, albeit always partially and aspectually. But...controversy crops up as soon as we start to ask for the details. (1999–2000, 6–7)

M. E. Turkis II, *The Metaphysics of Michael Polanyi*,
https://doi.org/10.1007/978-3-031-41211-0_3

We will survey these controversies in due order. First, I wish to briefly trace a clearly discernible through-line of what I will call Copernican realism running through all of Polanyi's philosophical works. This Copernican realist orientation is a variety of metaphysical realism that includes both scientific and epistemological realisms within its ambit. Polanyi does not treat these separately, however, but rather handles the latter two as logical entailments of his overarching metaphysical realism, and I shall adopt the same procedure here. I dub his position "Copernican" because it affirms that reality can be approached by both philosophical metaphysics and by empirical study, which is precisely the way Polanyi characterizes Copernicus's position, as we will see.

Polanyi marks out this realist position in the first section of *Science, Faith, and Society* (1946), returning to it as an explicit focus late in his career in his Duke and Wesleyan lectures (1964 and 1965a , respectively). He refined the ideas presented in these lectures and published them as "Science and Reality" in 1967. In this section I propose to distill the most salient and incontrovertible elements of his realism before reviewing briefly some of the disputes arising therefrom. From there we will proceed to examine what Polanyi called the ontological aspect of his theory of tacit knowledge as well as his often vague and cryptic references to fields and originating principles.

As a successful physical chemist, Polanyi's engagement with metaphysics is framed by his experience as a scientist and discoverer. He is adamant that science as actually practiced depends upon an assumed realist metaphysics that allows for scientists (and others) to make meaningful contact with reality by means of the verification and validation outlined in the last chapter. This is implicitly the case, he argues, whether or not scientists explicitly disavow such metaphysical commitments (such disavowals are thus examples of the bad faith that Cannon points out is part and parcel of the critical project). Thus scientific conclusions can encapsulate meaningful truths about reality and do not merely chart out mathematical relationships between observed phenomena, as the positivists who were Polanyi's contemporaries would argue.

Yet Polanyi's is not a naïve realism. In "Science and Reality" he proclaims that his intention is to.

> ...bring back the idea of reality and place it at the centre of a theory of scientific enquiry. The resurrected idea of reality will, admittedly, look different from its departed ancestor. Instead of being the clear and firm ground underlying all appearances, it will turn out to be known only vaguely, with an unlimited range of unspecifiable expectations attached to it. (1967, 177)

This resurrected reality is summed up in the introduction Polanyi wrote in 1963 for a new edition of *Science, Faith, and Society*:

> We meet here with a new definition of reality. Real is that which is expected to reveal itself indeterminately in the future. Hence an explicit statement can bear on reality only by virtue of the tacit coefficient associated with it. This conception of reality and of the tacit knowing of reality underlies all my writings. (6)

Despites the vagueness inherent in our explicit awareness of our tacitly rooted contact with reality, Polanyi argues that without such a metaphysical commitment, we cannot fruitfully pursue science: "It is impossible," he argues, "to pursue science without believing that it can discover reality" (1965a, 8).

In "Science and Reality" and his Duke and Wesleyan lectures Polanyi frames his position as a form of Copernicanism which, on his account represents a middle way between medieval scholasticism on the one hand and positivism on the other.

As Polanyi tells it, the general view of the scholastics was that contact with underlying reality could only be made by the practice of philosophy through reasoning about first principles. Empirical study was held to be beneficial, but could purportedly yield no knowledge of reality, circumscribed, as it was held to be, within the bounds of practical application. What Polanyi calls the medieval view, then, affirmed that reality was accessible, but only philosophically and *not* empirically.

The positivist position, also anathema for Polanyi, is that no contact with any sort of reality underlying phenomena is possible philosophically or empirically, and that any claim to the contrary is empty metaphysical speculation which ought to be got rid of, the sooner the better. In the controversy surrounding Copernicus, critics such as Osiander made proto-positivist arguments by holding that Copernicus had discovered a useful calculating tool for phenomena, and not a truth about reality. These arguments do not qualify as full-blown positivism since thinkers such as Osiander affirmed that contact with reality was possible in philosophical reflection, yet later positivists recognized their own position as prefigured by these medieval critics, although thoroughgoing positivists went further by denying philosophical reflection's ability to pierce the veil of phenomena.

Polanyi argues that Copernicus agrees with the medieval thinkers that contact with reality can be made, and that philosophy of first principles *is* a method capable of such contact. He (Copernicus) disagrees with them insofar as he believes that empirical study *is also* capable of such contact. We might call this Copernican position inclusive insofar as it affirms our ability to penetrate the veil of phenomena by both abstract philosophical and concrete empirical means. It is, again, the position Polanyi takes as his own, claiming that it is functionally and implicitly necessary for the pursuit of modern science, though such a realist metaphysics has often been explicitly disavowed. Polanyi is clear about his Copernican realism in his first Duke lecture, where he states:

> In this three-cornered dispute I shall side with the Copernicans; I agree with the positivists in pointing out that the Copernicans were making metaphysical claims, but I side against them by agreeing with the medievalists, that such claims are not empty but, on the contrary, essential to all true knowledge. Seen in this light, the medievalists were right, in recognizing empirical science as encroaching on their domain, but wrong in contesting that science was competent to do so. The Copernicans were right in every respect. (1964, 3–4)

Polanyi describes the conception of reality at play in this approach by explaining that

> What Copernicus believed of this system was what we all mean by saying that a thing is real and not a mere figment of the mind. What we mean is that the thing will not dissolve like a dream, but that, in some ways it will

> yet manifest its existence, inexhaustibly, in the future. For it is there, whether we believe it or not, independently of us, and hence never fully predictable in its consequences. (1967, 191)

Thus Polanyi sees his resurrected conception of reality as an extension of the Copernican revolution which, by means of its embrace of multiple methodologies, allows us to "abandon the cruder anthropocentrism of our senses – but only in favour of a more ambitious anthropocentrism of our reason" (*PK* 4). Reason thus liberated assimilates empirical observations into rationally coherent theoretical structures. He notes that

> ...a theory which we acclaim as rational in itself is thereby accredited with prophetic powers. We accept it in the hope of making contact with reality; so that, being really true, our theory may yet show forth its truth through future centuries in ways undreamed of by its authors. (Ibid., 5)

Similarly, in "Science and Reality" he explains that

> This defines reality and truth. *If anything is believed to be capable of a largely indeterminate range of future manifestations, it is thus believed to be real.* A statement about nature is believed to be true if it is believed to disclose an aspect of something real in nature. A true physical theory is therefore believed to be no mere mathematical relation between observed data, *but to represent an aspect of reality, which may yet manifest itself inexhaustibly in the future.* (1967, 191 emphasis added)

The italicized lines above are representative of Polanyi's definition of reality as independent and capable of surprising us in the future. Esther Meek has dubbed this "the IFM Effect," since on Polanyi's account "the knower's successful contact with reality on any particular occasion is accompanied by the intimation of *i*ndeterminate *f*uture *m*anifestations" (2017, 77 emphasis added). As I have already noted, I take Polanyi's definition of the real to be a form of the Eleatic Principle (see Plato 1997, *Sophist* 247d–e) whereby an entity's ontological standing is guaranteed by its causal power or efficacy in the world around it. The language of indefinite future manifestations is one way of keeping track of such causal power.

In addition to the IFM Effect, acknowledged nearly universally by Polanyi scholars as the crux of Polanyi's theory of the real, Walt Gulick draws attention to two additional elements of Polanyi's realism. The first is the criterion of independent existence, common to all realisms as traditionally conceived. The second is what Gulick calls the "significance criterion," (1999–2000, 28) since Polanyi holds that those entities capable of invoking deeper and richer fields of meaning are more real and hence higher on the hierarchical metaphysics he espouses. If, however, one rejects (as I do) the ontological distinction between the real and meaningful urged by Gulick, then the significance criterion simply becomes a way of keeping track of subgenres of IFMs.

In the case of the Copernican theory, Polanyi describes such intimations of indefinite future manifestations and the rich potential for meaning in the following way:

> We can ask then why the general appearance of the heliocentric system made Copernicus and his followers believe that it was real – why its close coherence, its intellectual harmonies had such power to convince them of its reality. And to this we reply that the existence of a harmonious order is a denial of randomness, and order and randomness are mutually exclusive. Moreover, anything that is random is meaningless, while anything that is orderly is significant. (1967, 191)

This emphasis on harmony and order is of a piece with Polanyi's discussions of the role of beauty in the scientific enterprise and in our knowing in general, a topic that formed a part of our earlier exploration of his epistemology. Indeed, Polanyi immediately ties such questions of order, meaning, and reality to the core of his epistemology – the personal and tacit coefficients of all knowledge:

> To distinguish meaningful patterns from random aggregates is therefore to exercise our power for recognising reality. Our capacity for discerning meaningful aggregates, as distinct from chance aggregates, is an ultimate power of our personal judgment. It can be aided by explicit argument but never determined by it: our final decision will always remain tacit. (1967, 191)

As we have already noted Polanyi's theory of knowledge is developed in tandem with a phenomenology of the heuristics of discovery. We are now beginning to see the way that for Polanyi this epistemology is bound up not only with such a phenomenology, but also an implied ontology and a commitment to a metaphysical realism which is not naïve. This ontology will be further explored below in chapters on the ontological aspect of tacit knowing and the metaphysics of comprehensive entities. Polanyi weaves this phenomenology into his discussion of Copernican realism as follows:

> To see a good problem is to see something hidden and yet accessible. This is done by integrating some raw experiences into clues pointing to a possible gap in our knowledge. To undertake a problem is to commit oneself to the belief that you can fill in this gap and make thereby a new contact with reality. Such a commitment must be passionate; a problem which does not worry us, that does not excite us, is not a problem; it does not exist. Evidence is cast up only by a surmise filled with its own particular hope and fervently bent on fulfilling this hope. Without such passionate commitment no supporting evidence will emerge, nor failure to find such evidence be felt; no conclusions will be drawn and tested – no quest will take place. (1967, 195)

This passage highlights the tacit foreknowledge that is constitutive of discovery and learning. Again, it differs from the medieval view as understood by Polanyi and outlined above insofar as it affirms that contact with reality is possible by both philosophical and empirical methods. It furthermore rejects the "false conception of truth and reality" assumed by "the anti-metaphysical critique of science" launched by positivism (1967, 195). Polanyi's ultimate goal here is thus to reestablish a Copernican realist metaphysics on the grounds provided by his epistemology, as is made clear when he concludes,

> Once the recognition of anticipatory powers in science establishes a conception of reality transcending tangible things, we might be able generally to acknowledge higher entities, intangible and yet real – as real as matter and yet meaningful. We shall recognise thus a cosmic hierarchy in which man has once more his own place. (Ibid.)

2 Controversies Surrounding Polanyi's Realism

Having thus established my own understanding of Polanyi as a Copernican realist, it is time to survey some of the readings put forward by other scholars of Polanyi's work.

Phil Mullins, for instance, perceptively points out that Polanyi's conception of the cosmic hierarchy "can be succinctly summarized by saying that those things that are real are preeminently signifiers" (1997). This is a strong gesture in the direction of my own conclusions about Polanyi's metaphysics, which I believe center around the affirmation that *to be is to mean*. In this scheme, "some real things, such as challenging problems and minds, are more deeply real in that they are richer signs which hold the potential to generate a whole field of meaning or future significance not presently recognized" (Ibid.). Mullins argues that Polanyi advocates a "participative realism" which is based upon his conception of comprehensive entities, a subject we will soon examine in more detail. On Mullins' reading, real things as signifiers "are the comprehensive entities which are the result of indwelling and integration. [Thus all] realities are the end product of tacit integrations rather than merely the external cause of such integrations," while "claims about the independence and prior existence of realities are inferences introduced to account for the compelling coherence of our integrations" (Ibid.). This interpretation of Polanyi's realism implies, crucially, that a Polanyian ontology is flat insofar as it is made up of comprehensive entities, but that this flat ontology nestles within a cosmic hierarchy. As will be seen, this is my own position as well, though I plan to develop it later in a distinctly Platonist direction not affirmed by Mullins' writings.

Not all interpreters of Polanyi are as approving of Polanyi's unified vision of reality as Mullins, however. Stephania Jha (1999–2000), operating within the tradition of mainstream analytic philosophy, is suspicious of Polanyi's jump from knowing to being and argues that his ontology is an unstable mixture of realism and anti-realism. She goes on to say that Polanyi's is a "'modified rational realism,' if labels from mainstream [analytic] philosophy are insisted upon" (Ibid., 21).

Walt Gulick (1999–2000), following Harry Prosch (1972), argues forcefully that Polanyi's realism ought to be developed by separating the noosphere (the realm of human-constructed meanings) from the realm of the real (defined as the empirical). In general, Gulick, wishes to save "reality" for reference only to those entities which are entirely empirical or which exhibit a mixture of meaning and the empirical (by which he means those entities which are themselves human constructed meanings, but which gives rise to empirical components). He thinks the all-encompassing nature of Polanyi's realism as read by Mullins (1997, 1999–2000) is "unsatisfactory because it has no discriminatory power" (Gulick 1999–2000, 29).

While it is clear that Gulick's intention is not to demote meaning in any way, other commentators such as Mullins, Cannon, and Sanders worry that he may be setting the stage for just such a demotion, since by cordoning off the realm of meaning from the realm of the "existent" his proposal would undo the unifying work accomplished by Polanyi's contributions and actually feed into relativism insofar as, without the imposition of a certain range of value judgments (nonexistent since non-empirical, yet meaningful on Gulick's account) this position is arguably quite close to emotivism or a range of similar positions that, according to Polanyi, precede moral inversion. Gulick's response would likely be to mine Polanyi's late work, *Meaning* (1975), for support for his proposed divisions, however the handling in *Meaning* of the division between what Gulick calls the realm of the real and the realm of the meaningful is contested among Polanyi scholars since the book, which Polanyi could not finish himself due to the onset of senility, was co-written with Prosch, who himself advocated such a divided approach to ontology. Interpretations of Polanyi's realism split to a significant extent along the fault line of the *Meaning* controversy, with Gulick and Prosch on the side favoring an ontological distinction between the empirical and meaningful and Mullins, Cannon, and Sanders opposing such a distinction. My own understanding of Polanyi's Copernican realism aligns more closely with the group opposing the distinction.

Dale Cannon points out that Prosch and Gulick seem to "hold to an idealist interpretation of Polanyi's account of cultural meanings but to a realist interpretation of Polanyi's account of natural science" (1999–2000,

52). This does not sit well with him, however, and, citing passages from PK 201–202 that give equal ontological weight to verification (for empirical questions) and validation (for questions of cultural meaning) as well as on beauty as a token of reality, he remarks that Polanyi

> does not hesitate to speak of validating realities here, and he clearly identifies art and mathematics as both concerned with reality. So what are we to make of it? …A possibility that should be considered is that he is adumbrating a novel but coherent philosophical position that is neither realist nor idealist as traditionally conceived, a comprehensive (post-critical) realist position that concedes some truth to each… (Ibid., 52–53)

Cannon brings to the fore Polanyi's emphatic insistence that no contact with reality, whether empirical or cultural, is possible without passionate personal commitment and concludes that Polanyi is a realist, but a realist who does not fit neatly in analytic categories given that he does not "presume[e] to speak of noumenal reality, reality in itself, noncommittally" since, "considered apart from the framework of commitment that Polanyi sets out, to presume to speak of noumenal reality non-committally, as if it could be discussed from a non-committal standpoint…is to beg the question of what is reality and how it is to be determined…" (1999–2000, 54).

Meanwhile, John Puddefoot calls Polanyi's approach "resonance realism," arguing that what makes it distinctive is "the faith…that justifies our submission to [reality] as a source of unlimited richness, and our consequent willingness to identify the richness of our conceptualizations with the richness of the reality that they arise from" (1993–1994, 30). Thus, without attempting to apply a correspondence theory of truth to verify or validate our concepts, "we allow the richness of those concepts themselves, as measured by the resonances they evoke from the world and in our own minds, to be their authentication" (Ibid.). For Puddefoot, traditional metaphysical realisms are "foot-stamping totalitarianism[s] in disguise" (1999–2000, 70), thus he wants to clearly cordon Polanyi's realism off from such a label, although most interpreters of Polanyi would disagree with this stance on metaphysical realism.[1]

Andy F. Sanders holds that Polanyi's is a "comprehensive realism" (1999–2000, 89) that affirms the reality of both the empirical and meaningful and includes both value realism (with regard to the transcendentals) and theological realism. He takes issue with the ontological distinction between the empirical and the meaningful urged by Gulick and argues that he (Gulick) is, despite his own best efforts, re-opening the fact-value distinction that Polanyi's realism seeks to undermine (Ibid., 91).

Esther Meek, whose coinage (the IFM effect) we have already invoked, notes Polanyi's "commitment to…epistemological and metaphysical realism" and points out that it is "less a pragmatic realism and more something like a fiduciary realism" (Meek 2017, 203). Ultimately, she sees Polanyi's definition of reality as drawing us into an "exuberant realism" in which the IFM effect generates a one-to-infinity correspondence, rather than a paltry one-to-one relationship.[2] In her view this is because

> The heart of Polanyi's own realism lies in his vision of reality as that which manifests itself indeterminately in the future, and thus of any successful contact with it being indicated by an unspecifiable sense of indeterminate future manifestations. There is a three-dimensionality, a fecundity, an indeterminacy, a surprise, and an ongoing transformativeness about this that, itself true to this Polanyian signature of the real, more than matches and confirms; in contrast, it outstrips and redefines, redrawing the very question of realism itself. (Meek 2017, 203)

Before moving on, we would do well to pause for a moment over the IFM test of reality. While many if not most Polanyian thinkers take IFMs to be central to a proper understanding of reality, this view is not uncontested.

Gulick, for example, pushing back against the "meaning/reality holism" embraced by Mullins, argues that "a significant cause of disagreements

[1] See for example Gulick (1999–2000), Sanders (1999–2000), and Cannon (1999–2000), all of whom attribute to Polanyi some form of metaphysical realism and explicitly deny the inherent totalitarianism of such a position.

[2] Esther Meek, email message to the author, October 27, 2020.

concerning Polanyi's notion of reality is that the IFM Effect applies both to independently existing things and significant things which may be mind-dependent" (Gulick 1999–2000, 27–28). He insists that "the IFM Effect is better understood as a measure of important experiences of meaning than as a marker of the real" (Ibid.)

Gulick correctly notes that the Eleatic Principle "has much in common with the IFM Effect" (Ibid., 29).[3] He argues that Plato and Polanyi's joint affirmation of the Eleatic Principle entails an understanding of reality – "the sum of things" (1997, *Sophist* 249D), including both the concrete and abstract, particulars and Forms – which is in his view is lacking discriminatory power (Gulick 1999–2000, 29).

Gulick is spot-on when he links the IFM Effect to a Platonic holism and the Eleatic Principle, but misses the mark when he claims that an ontology that is flat in the sense that it grants real status to both the empirical and meaningful lacks discriminatory power. Algebra and Geometry are both Mathematical, yet easily and precisely distinguishable for all that. On a terminological level, then, *empirical* and *meaningful* do just fine when distinguishing between different subclasses of real entities, as Gulick's own, lucid prose demonstrates quite well.

But as should already be apparent, my own inclination is not merely to approve of the meaning/reality holism that Polanyi has in common with Plato, but rather to lean into that commonality with some force in order to unearth metaphysical resources that Polanyi needs and that Plato can provide in order to shore up the linkage between knowing and being that Polanyi repeatedly affirms. Before jumping with both feet into these metaphysical details, however, we will examine Polanyi's Copernican realism in the light of three contemporary realist movements.

---

[3] Similarly, Jha notes that Polanyi "drew on a Platonic-Aristotelian kind of formism relying on the general notion of similarity. His numerous analogies are built on the notion of similarity of some features, where to explain conceptions outside of or straddling frameworks, he relies on either a series of free-standing or of overlapping analogies" (1999–2000, 20). She does not seem entirely comfortable with this aspect of his thought. With regard to the analogical element of Polanyi, Phil Mullins (email message to the author, September 1, 2020, article forthcoming) has argued that Dorothy Emmett's work on analogy in metaphysics likely played an influential role in Polanyi's philosophical development.

3 Polanyi and Three Contemporary Realisms: Ontic Structural Realism, Object-Oriented Ontology and Neo-Aristotelian Hylomorphism

3.1 Ontic Structural Realism

In the years since the above critical work was produced, several developments and novel trends in realist philosophy have emerged which are relevant to our discussion. In the analytic world, a fair share of scientific realism has gone in the direction of structural realism, which, in the face of the underdetermination argument (which aims at undermining our ontological commitment to unobservable entities postulated by our best scientific theories) and the pessimistic meta-induction (which aims at undermining our commitment to the best, current scientific theories in light of the rejection of the best theories of the past) emphasizes the structural consistency from one scientific theory to subsequent theories that replace it, arguing that there is "continuity or accumulation in the shift, but the continuity is one of form or structure, not of content" (Worrall 1989, 117).

Thus, says the structural realist, we ought not commit ourselves to unobservable entities postulated by our best theories, but we should not be anti-realists with regard to science either, since we can, on this account, safely commit ourselves to the mathematical and structural content of such theories. The position is thus one of semi-positivism, since we are urged by this theory to remain agnostic about the underlying entities themselves, though there is typically a more robust metaphysical commitment to the ontology of the structure itself than traditional positivists would have countenanced; it seems to be more than merely a useful calculating device, though the language surrounding this is often ambiguous.[4]

James Ladyman argues for a more thoroughly metaphysical version called ontic structural realism (OSR) which posits that there are (or may

[4] See Ladyman (1998) on whether structural realism should be taken as a metaphysical or merely epistemological proposal.

well be) no "things" or underlying entities and that in light of this, the only thing to which we can confidently grant ontological standing is structure. He clarifies that

> When we go on to deny that, strictly speaking, there are "things," we will mean to deny that in the material world as represented by the currently accepted scientific structures, individual objects have any distinctive status. Some real patterns, we will argue, behave like things, traditionally conceived, while others behave like traditional instances of events and processes. In a PNC-motivated metaphysic these distinctions lose all significance except purely practical, book-keeping, significance for human beings in certain sorts of special circumstances. From the metaphysical point of view, what exist are just real patterns. (Ladyman et al. 2007, Location 2009)

OSR, at least as propounded by Ladyman, advocates a strictly limited metaphysics whose only legitimate job consists of correlating two or more scientific hypotheses in the ways deemed appropriate by the Principle of Naturalistic Closure (the "PNC" referenced above):

> Any new metaphysical claim that is to be taken seriously at time *t* should be motivated by, and only by, the service it would perform, if true, in showing how two or more specific scientific hypotheses, at least one of which is drawn from fundamental physics, jointly explain more than the sum of what is explained by the two hypotheses taken separately, where this is interpreted by reference to the following terminological stipulations:
>
> Stipulation: A "scientific hypothesis" is understood as an hypothesis that is taken seriously by institutionally bona fide science at *t*.
>
> Stipulation: A "specific scientific hypothesis" is one that has been directly investigated and confirmed by institutionally bona fide scientific activity prior to *t* or is one that might be investigated at or after *t*...where this activity is part of an objective research project fundable by a bona fide scientific research funding body. (Ibid. Location 671)

With this brief exposition in place, we can proceed to compare OSR to Polanyi's approach, fleshing out our understanding of OSR along the way.

Polanyi's realism is indubitably at odds with OSR as propounded by Ladyman in significant ways, however there is also common ground that could provide a basis for interesting discussion between the two positions with regard to what OSR calls the scale-relative ontology of real patterns (following Daniel Dennett, 1991) and Polanyi's comprehensive entities as well as on how to blend realism and verificationism.

Polanyi would fiercely oppose the openly scientistic orientation of the OSR project (Chap. 1 of Ladyman and Ross's book is entitled "In Defense of Scientism"). He would also surely reject the limited scope given to metaphysics that is part and parcel of this scientism and which is clearly embodied in the PNC described above. Polanyi's Copernican realism affirms empirical and scientific proceedings as able to teach us about reality, but it also maintains that metaphysics is a valid method for contacting the real. Thus, to the extent that it hamstrings philosophy in its quest for truth, the PNC, for Polanyi, could fairly be considered an extension of "the anti-metaphysical critique of science" (1967, 195) he set out to overcome. For Polanyi, philosophy and science are partners in investigating the real. For OSR, philosophy ought to be hemmed in by current scientific consensus.

Polanyi would also, I believe, vigorously critique OSR's dependence on an abstract and reified understanding of "science." Science and scientists are mentioned often in Ladyman's texts, but these terms take on at once a mythical force and a curious vacuity. Science *per se* is revered as the Olympian height from which all thought that aspires to be taken seriously must come. Scientists function as a sort of Olympian pantheon – superhumans with mysterious abilities surpassing those of the mere mortals who dwell at the benighted level of medium size dry goods. Yet these gods are strangely empty figures, almost algebraic variables devoid of content. Nowhere do Ladyman or his colleagues address how a scientist can be made out of flesh and blood, the typical human material.[5] This would be a major point of contention for Polanyi. His thick description of the phenomenology of scientific discovery, along with the cogent epis-

[5] Ladyman and Ross might respond that their view of science is based on a sort of Peircean pragmatism or instrumentalism, yet Peirce is much more insightful and honest when looking at the workings of science in terms of the concrete operations of scientists. Much of this work of Peirce's would, I believe, violate the PNC, though such an argument lies beyond the scope of this work.

temological arguments he provides – arguments which flow from his own experience of scientific discovery – are a much surer guide to the ways in which science arises from the mortal coil and is thus always tethered to it. Furthermore, Polanyi would insist that the operation of science requires a more inclusive metaphysical view (his Copernican realism), whether tacitly or explicitly, than is allowed for by the PNC as stipulated by theorists of OSR.

To get a better sense of where the conflict between these two realisms lies, consider the following. Ladyman explains that, "in order to be able to refer to particular real patterns… without relying on the conventional distinctions between objects, events, and processes," one depends on.

> …an ostensive operation – that is, by "pointing at it." This is intended as evocative talk for operations of fixing, stabilizing, and maintaining salience of some data from one measurement operation to another. (So, think of "pointing" as meaning "directing a measurement apparatus.") (Ladyman et al. 2007, Location 2013)

The operation of "pointing at" something is not unpacked or explained other than to define it in sterile mechanical terms ("directing a measurement apparatus"). The operation is presented as wholly impersonal and uncentered. There is pointing, but nobody who points. The measurement apparatus is mysterious. Whence come the units of measurement and why point to this rather than that? Doubtless Ladyman would recur to the institutions of science as the proper source for answers to all of these, however there is a clear and vicious circularity here, since the operation of scientific institutions depends entirely on the "pointing" activities not only of scientists, but also additional personnel who are non-scientists as well as a society made up of individuals and families in which science itself can unfold. The role of Eros – the intrinsic interest that spurs on scientific research and which is unavailable noncommittally – as the central motivating force for all pointing activity is completely ignored.

Much truer to life and better explanatorily is Polanyi's theory of personal knowledge – persons as centers of integrative achievement, active in the world, motivated by passion to tacitly integrate the subsidiary particulars whose joint meaning is the comprehensive entity to which they contribute. "Pointing" or "directing measurement apparatuses" does not come near to doing justice to the noetic realities of everyday life, let alone scientific practice. The OSR theorist might respond that scientific methods somehow function to epistemically rise above the fray of everyday life, thus throwing off the mortal coil, however the considerable weight of Polanyi's epistemological arguments militates against this claim: After all, scientists are people, too. They, as a group of connoisseurs within the larger community, select topics for research, generate hypotheses, select from among these the hypotheses that seem most likely to be fruitful, search for funding, design and build equipment, suffer setbacks in their research, and, significantly, *interpret* observed data intuitively (see *SFS* 15).

Ladyman and his colleagues might respond by arguing that

> As collective constructions, the institutional filters of science need not mirror or just be extensions of individual cognitive capacities and organizing heuristics. They have shown themselves to have a truth-tracking power – partly thanks to mathematics – that bootstraps the process of scientific learning beyond the capacities of individual minds. (Ladyman et al. 2007, Location 4994)

Yet Polanyi's insider account of the institution of science is far thicker than this thin assertion of the power of bootstrapping. Polanyi would agree that the enterprise as a whole does indeed transcend the cognitive and heuristic activities of the individual scientists involved, but would also show how the process by which individual cognition and heuristics is transcended is itself constituted by precisely the same cognitive and heuristic resources found at the individual level.

For instance, when research, based on the work done in a laboratory under the direction of Scientist A is sent out for peer review prior to publishing and it is reviewed by fellow experts Scientists B, C, and D, there is no escape from the tacitly based personal and fiduciary co-efficients of judgement, as these are still exercised by B, C, and D as they carry out their evaluations of the work in question. Even if they base their critique of the work on mathematical grounds, or frame their critique in mathematical terms, they still do so by means of committed, personal acts of judgement which they carry out with universal intent. If B and C call for A's work to be published, while D expresses reservations, then Editor Z steps in and exercises her heuristic powers. Will she send the work out to another reviewer? Will she take B and C's approval as sufficient since she knows that D is often overly fastidious or may have some bone to pick with the work that Z considers inappropriate?

The point here is that no mysterious, non-tacitly-rooted mathematical bootstrapping is possible, since in the final analysis, individual scientists making up specialized sub-groups will either commit epistemically or not to new theories by means of these same heuristic devices. This effect is only amplified when we consider, as Polanyi clearly explains time and again, that the partially overlapping areas of expertise of specialized sub-groups logically require a sort of trust or faith that adjacent and more far-flung sub-groups with other specialized areas of interest are keeping their own heuristic houses in order. Thus the potential failings of individuals in their heuristic endeavors can be reigned-in by scientific institutions, but individual heuristic commitment for which no non-circular justification can be given remains ubiquitous and inescapable since it too serves as the very basis for such reigning-in.

One further point ought to be made on the bootstrapping role of mathematics referenced above by Ladyman. (This point is doubly important given that for OSR, the reality that science deals in is, if not itself mathematical structure, then at the very least is tracked by its isomorphic or homomorphic relation to mathematical structure.) Polanyi observes that

> Never yet has a definite rule been laid down by which any particular mathematical function can be recognized, among the infinite number of those offering themselves for choice, as the one which expresses a natural law. (*SFS* 7)

This too is damning. Here the scientist must make use of her intuition to select the right mathematical function – the one that will, by means of the intuition's grasp of reality, generate the richest array of indefinite future manifestation. This success comes about in an unformalizable way, motivated by her passionate interest in the solution to the experimental problem, guided by tacit, subsidiary clues.

The problem for OSR is that scientists' dependence on the unformalizable and, at the explicit level, *ad hoc* use of tacit knowhow – inevitably interpreted by OSR as *intuition* since it is neither explicit nor formalizable – does not meet the requirements laid down by the PNC. Ironically, this means that on the OSR account, philosophers are bound by the PNC, while practicing scientists are not. The PNC is of course intended to prevent philosophers from making unfettered use of "intuition" in their work by tethering them to the findings of science, yet all the while it turns out that the findings of science rely precisely on the exercise of unformalizable connoisseurship and "tacit foreknowledge" (CI) within the laboratory itself in order to constitute the very consensus that the PNC depends on. Polanyi elaborates on the near constant need of a research scientist to make such discriminating, intuitive leaps and guesses:

> There is always the possibility that...a deviation may not affect the essential correctness of a proposition. The example of the periodic system and of the quantum theory of light both show how the objections raised by a contradiction to a theory may eventually be met not by abandoning it but rather by carrying it one step further: Any exception to a rule may thus conceivably involve, not its refutation, but its elucidation and hence the confirmation of its deeper meaning. The process of explaining away deviations is in fact quite indispensable to the daily routine of research. In my laboratory I find the laws of nature formally contradicted at every hour, but I explain this away by the assumption of experimental error. I know that this may cause me one day to explain away a fundamentally new phenomenon and to miss a great discovery. Such things have often happened in the history of science. Yet I shall continue to explain away my odd results, for if every anomaly observed in my laboratory were taken at its face value, research would instantly degenerate into a wild-goose chase after imaginary fundamental novelties. (*SFS* 17)

Here we see the centrality of Polanyi's affirmations that the "intuitive powers of the investigator are...dominant and decisive" (*PK* 130). In contrast to OSR theorists' distaste for the messiness of "intuition" Polanyi argues explicitly that it is an inextricable part of the scientific enterprise since researchers must shift their "confidence from intuition to computation and back again" (Ibid.) and this "alternation between the intuitive and the formal depends on tacit affirmations both at the beginning and at the end of each chain of formal reasoning" (Ibid., 130–131). Intuition for Polanyi thus refers to our reserves of tacit and explicit knowledge along with our unformalizable skills and our ability to tacitly identify relevant subsidiary details and then to integrate them, even in novel situations liable to give rise to scientific discovery. Thus

> There is a *residue of personal judgement* required in deciding — as the scientist eventually must — what weight to attach to any particular set of evidence in regard to the validity of a particular proposition. (*SFS* 17 emphasis added)

Personal judgement is, to use OSR's own phrasing, not PNC compatible. If science is ultimately dependent upon the exercise of such personal judgement, then neither is it PNC compatible, and thus the quasi-positivist branch attacks the root which nourishes it.[6] But this, of course, is precisely what Polanyi showed with regard to the explicit and more thorough-going positivism of his own time.

Despite all this, I suspect that some fruitful conversation might arise between Polanyian perspectives and OSR on the topic of its proposed scale-relative ontology of real patterns as well as on its blend of realism and verificationism.

Polanyi's conception of comprehensive entities, addressed ever so briefly in the chapter on epistemology and due for a more thorough examination shortly, most certainly has significant overlap with OSR's conception of real patterns, though I see Polanyi's account as superior, as will become clear.

OSR's terminology of real patterns is drawn from Daniel Dennett's 1991 essay of the same name. There are, from a Polanyian perspective, real benefits to using such a scale-relative terminology over more traditional scientific realist vocabularies that assume an ontological reduction

[6] I say "quasi-positivist" because given that OSR affirms the mathematical structure retained through the evolution of scientific theory as real (often treating it as interchangeable with nomological structure), and not merely a useful calculating tool, it is not positivist in a strict sense.

to the smallest entities postulated by current physics. Ladyman et al. follow Dennett in defining real patterns in information-theoretical terms:

> [Dennett] says, 'a pattern exists in some data – is real – if there is a description of the data that is more efficient than the bit map, whether or not anyone can concoct it' (1991, 34). Thus there are (presumably) real patterns in lifeless parts of the universe that no actual observer will ever reach, and further real patterns whose data points are before our eyes right now, but which no computer we can instantiate or design will ever marshal the energy to compact. (Ibid. Location 3373)

In Polanyian terms the metaphysical benefit of this approach over the general thrust of analytic realism is that it accords real ontological status to emergent entities. There is even a sense in which it moves further in this direction than Polanyi explicitly does, since it does not see the special sciences which track various sorts of real patterns at different scales as responsible for demonstrating the mechanisms by which such patterns emerge. By OSR's lights such emergent explanation is potential fair game, but it is not required for the justification of the tracking of a real pattern, as seen in the definition above.

On the other hand, Polanyians will want to leverage a critique of OSR's understanding of real patterns which is similar to the one launched against its impersonal conception of science above. The problem here is not so much that real patterns are defined in terms of information theory, but rather that the information theory in question is itself depersonalized in its tacit dependence on a computational theory of mind which not only puts the cart before the horse, but forgets the horse altogether.

I would argue that Polanyi's comprehensive entities can do all of the beneficial metaphysical work that real patterns do without falling into the trap of epistemic depersonalization, and without committing us too firmly to either side of the debate between OSR and traditional scientific realism over whether reality is best understood in terms of substance or relation.

We should recall that on Polanyi's account, comprehensive entities are the meaning toward which an unspecifiable array of tacitly-indwelt subsidiary particulars point, and that such comprehensive entities are comprehended through the tacit integrations of the knower. Comprehensive entities, however, are not always ontologically dependent on the epistemic activities of the knower, since they manifest their independent reality as indicated by their IFMs.

On the basis of this brief review, note first that comprehensive entities can be construed as either substance or relations, and if construed as the former, their substantiality *is due to their integrative relations*. We might even say that Polanyi's coinage is particularly potent precisely because it taps into both categories – "comprehensive" encapsulating the aspect of relationality and pointing toward the integrative nature of reality, and "entity" pointing to the substantiality and ontological impact that such relational integrations can have. Nonetheless, Polanyi's view is superior to OSR's real patterns ontology since it does not involve a screening out of the integrative role of personal centers of achievement as integral to human contact with reality.

Furthermore, as we will see in our in-depth discussion of comprehensive entities, this concept is equally applicable to cobblestones and cobblers, mountains and music, Shakespeare and Shylock. OSR's approach can handle all but the last in this list, Shylock, since he is a fictional character whose real existence, with the potential for influence beyond the bounds of literary formalism (say, in terms of both critiquing the tradition of European antisemitism and also perhaps of carrying on and fomenting that very tradition) is ruled out of bounds by the PNC.

In a certain, restricted sense, then, Polanyi's ontology is structurally flat, though within this flat structure a hierarchy of ascending levels of reality, distinguishable by means of their varying potency for indefinite future manifestations, can emerge. The Polanyian position which holds that reality is made up of comprehensive entities, then, is in important respects similar to OSR's claim that "the real patterns criterion of reality is the last word in ontology, and there is nothing more to the existence of a structure than what it takes for it to be a real pattern" (Ibid.). Again, however, it does OSR one better by avoiding the scientistic depersonalization characteristic of much analytic discourse.

Though it is not a word Polanyi uses explicitly, OSR's descriptive term "scale-relative"[7] is clearly applicable to comprehensive entities, epistemologically and ontologically, since we can shift our focus between subsidiary details (themselves comprehensive entities in their own right) and the comprehensive entities to which they point, and both sides of this equation are real. One thinks here of the stones arranged into words on the hillside mentioned by Polanyi (*PK* 33–34) – the message, the words, the letters, the stones, the chemical structures, etc. are all real, all comprehensive entities.

Furthermore, since proponents of OSR deny the illata/abstracta distinction and are explicitly anti-reductionist (believing that contemporary science militates against such reductionism),[8] they argue that their scale-relative ontology (made up of real patterns irreducible to any fundamental level of "things" to mechanistically "microbang") (Ladyman et al. 2007, Location 120) requires that they embrace a "rainforest realism:"

> ...of lush and leafy spaces rather than deserts, with science regularly revealing new thickets of canopy. Anyone is welcome to go on sharing Quine's aesthetic appreciation of deserts, but we think the facts now suggest that we must reconcile ourselves to life in the rainforest. (Ibid. Location 3910)

Thus OSR theorists share with Polanyi a putatively anti-reductionist stance and ontological scale-relativity. They are open to tracking a plethora of irreducible real patterns by means of the special sciences. Thus far the Polanyian can travel with them. Yet, due to the scientistic restrictions on metaphysics laid down by the PNC, values cannot be real patterns, since, as Ladyman puts it, "naturalized metaphysics and science cannot tell us what to value" (Veal 2012, 155). Values are thus ontologically demoted on OSR's account in a sort of implied emotivism, a result which is unacceptable for Polanyi's approach given that one of the ultimate aims

[7] Scale relativity is "...the idea that which terms of description and principles of individuation we use to track the world vary with the scale at which the world is measured...Here are some [...examples]: at the quantum scale there are no cats; at scales appropriate for astrophysics there are no mountains; and there are no cross-elasticities of demand in a two-person economy" (Ladyman et al. 2007, Location 3317).

[8] This anti-reductionist stance could theoretically change with future science – it is not an *a priori* matter, as anti-reductionism is for Polanyi).

of his project is to prevent moral inversion by providing epistemological and ontological resources justifying the place of moral values in human life.

Thinkers of the OSR school might well argue that Polanyi's terminology is hopelessly muddled and imprecise. Yet this would be due to their insistence on bowing down to the golden calf of "liberation from anthropocentric myths by [means of] formal articulation" to be achieved by "leaving natural language behind" (Ladyman et al. 2007, Location 3941). Polanyians will rightly respond that this linguistic pipe dream is built on the shoddy foundation of an objectivist epistemology which fails to account for the tacit, personal, fiduciary co-efficients of all knowing, including the scientific knowing that OSR looks to as a source of priestly guidance in its "frank scientism" (Harman 2010, 273).

Finally, we turn to OSR's blend of realism and verificationism. It should be clear by now in what ways OSR takes itself to be realist: the mathematical structure retained in the development of scientific theory (scale-relative real patterns) is real and is either identical to or so closely related as to be indistinguishable from the nomological structure it "describes."[9] The verificationism in question is a

> ...non-positivist version of verificationism...that is universally respected by the institutional practices of science...consist[ing] in two claims. First, no hypothesis that the approximately consensual current scientific picture declares to be beyond our capacity to investigate should be taken seriously. Second, any metaphysical hypothesis that is to be taken seriously should have some identifiable bearing on the relationship between at least two relatively specific hypotheses that are either regarded as confirmed by institutionally bona fide current science or are regarded as motivated and in principle confirmable by such science. (Ladyman et al. 2007, Location 529)

[9] "The idea of structure that I have in mind is that of modal structure – nomological structure – rather than mathematical structure. However, it might be that the necessity of mathematics and the physical necessity of laws coincide, and the history of physics has progressively undermined the distinction between mathematical and physical necessity. We do not advocate Pythagoreanism or Platonism, because they contrast mathematical or formal existence with material being. Our metaphysics is directly motivated, just as we say all metaphysics should be, by the actual content of physics, and our review of the latter significantly undermines the distinction between mathematical and physical-nomological structure. However, if mathematical structure does not coincide with physical structure then the relationship between them is surely that of isomorphism (or something similar but weaker such as homomorphism)." (James Ladyman, quoted in Veal 2012, 165)

If this form of verificationism were meant simply to define some useful rules of thumb for appropriate scientific hypothesis generation, then I think Polanyi would approve. Since, however, it is not merely a guide to scientific practice, but is rather a metaphysical dogma tightly bound up with OSR's conception of the PNC, it is in direct opposition to Polanyi's Copernican realism, which affirms both empirical and philosophical (especially with regards to first principles) approaches to reality. From a Polanyian perspective, then, OSR's is a closed and dogmatic version of verificationism which artificially restricts the appropriate scope for philosophical investigation on spurious grounds.

Nonetheless, Polanyi's approach does seek to blend realism and verificationism as well, though his verificationism is open, intuitive, and non-dogmatic – consisting, as it does, of the IFM test of reality along with a non-scientistic understanding of the operations of scientific institutions, to say nothing of its pairing with validation and authenticity (*PK* 202), which are anathema to OSR unless they can be reinterpreted through its own proprietary verificationist lens. On Polanyi's reading these institutions collectively harness, direct, and refine what is fundamentally a tacit, imaginative, and intuitive process that unfolds in largely unformalizable ways dependent upon the tacit foreknowledge of the scientific discoverers and the personal, heuristic commitments of colleagues, peer-reviewers, funders, etc. The entirety of this epistemic process, then, according to Polanyi, requires metaphysical resources that run far beyond the bounds of the PNC. He speaks

> ...repeatedly of our powers to perceive a coherence bearing on reality, with its hidden future manifestations, and...also a more specific knowledge of hidden coherence: the kind of foreknowledge we have in a problem. And we know that the scientist's imagination sees problems, produces hunches, and, stirred by these anticipations, pursues his quest towards them. We may say that this quest is guided at every stage by feelings of a deepening coherence, of feelings that have a fair chance of proving right. We may recognise in them the working of a dynamic intuition: in other words, the power which achieves tacit knowledge. (*CI* 6–7)

To conclude, then, Polanyi's realism does feature some noteworthy overlap with OSR. They both actively oppose reductionism on the basis of scale-relative ontological commitments. OSR's understanding of reality as composed of real patterns explicitly denies their "substantiality," traditionally conceived, and holds that reality may very well be composed of relations without relata. Meanwhile, Polanyi's conception of comprehensive entities is not overly or explicitly dependent on traditional conceptions of substance, though his view is, as I will argue more fully later on, flexible with regard to distinctions between particular substantive entities, events, and relations, and is thus open to the substantiality of comprehensive entities in general. Furthermore, OSR and Polanyian realism alike move beyond naïve realisms by blending scientific realism and verificationism, though, as we have seen, there are important differences in their understanding of the proper bounds of both of those categories.

On the other hand, Polanyi's realism is not constituted around a demand for closure under physics and rejects outright the artificial boundaries for metaphysics that OSR stipulates by means of its Principle of Natural Closure. The metaphysical task of integrating the various special sciences with physics is, for Polanyi, an important and worthwhile activity, but it does not define the whole of metaphysics properly conceived. Philosophy is rather competent to deal in first principles and in fact his view affirms the necessity of metaphysical first principles which are not themselves susceptible to scientific analysis for the workings of science to rationally proceed. Furthermore, the analytic pipe dream of abandoning natural language for perfect formalizability in the hope of escaping the "parochialism" of humanity is, on a Polanyian account, a dangerous and preposterously hubristic red herring.

While Polanyi would, I think, see OSR's open rejection of reductionism as an indubitable improvement over reductionistic forms of analytic realism, he would see the scientistic rationale for this rejection as worse than suspect, given that the current scientific consensus upon which OSR's metaphysical judgement against reductionism depends might change, thus potentially bringing reductionism back into the proper purview of OSR, by its own PNC lights. In this way, then, OSR remains a potentially dangerous metaphysics since it does not definitively guard against the slippery slope to the unreality of values and moral inversion at the level of first principles.

3.2 Object-Oriented Ontology

Our focus shifts now to another, very distinct contemporary realism: Object-Oriented Ontology (OOO hereafter). OOO's leading theorist is Graham Harman, though he has been joined by thinkers such as Levi Bryant, Timothy Morton, Ian Bogost, and others. OOO forms one part of a resurgence of interest in realism in the sphere of continental (or post-continental) philosophy which has been called speculative realism. This larger group – comprised of figures such as Quentin Meillassoux, Ray Brassier, Manuel Delanda, Ian Hamilton Grant, etc. – is united in its opposition to what it takes to be post-Kantian philosophy's undue emphasis on the centrality and finitude of human thought and its access to reality, which thinkers in this movement often call *correlationism*. In Harman's words correlationism is characterized by some form of the following feedback loop: "to think an unthought X is to turn it into an X that is thought" (Harman 2010, 785).[10] Such a position is, he thinks, an inherently idealist trap, though those who take it typically try "to avoid charges of idealism by appealing to some excess beyond what is currently formalized" (Ibid.).

In what follows we will engage in a brief exposition of Harman's own proposed metaphysics before bringing it into conversation with Polanyi's. I will argue that Harman's OOO has many more constructive points of contact with Polanyi's project than does Ladyman's OSR. Furthermore, given that Polanyi never provides anything close to a systematic metaphysics, those interested in fleshing out more fully a metaphysics compatible with Polanyi's epistemology may find many useful items in the OOO

[10] Quentin Meillasoux states "By 'correlation' we mean the idea according to which we only ever have access to the correlation between thinking and being, and never to either term considered apart from the other. We will henceforth call correlationism any current of thought which maintains the unsurpassable character of the correlation so defined. Consequently, it becomes possible to say that every philosophy which disavows naïve realism has become a variant of correlationism" (2006, 5).

toolbox. This task will require a considerable amount of exposition, since OOO lies farther afield from traditional realisms than does OSR, or, for that matter than does Polanyi's Copernican Realism.

To a significant extent, Harman is interested in rehabilitating a theory of substance with classical antecedents, largely on the basis of insights generated by twentieth century philosophers who sought to surmount the correlationist obstacles thrown up by Kant and his followers. He notes that even though both Husserl and Heidegger are themselves working within correlationism,

> ...both silently raise objects to the starring role [of the human subject], each in a different manner. While Husserl bases his system on intentional or ideal objects (which [he] will rechristen sensual objects), Heidegger restores real objects to philosophy through his famous tool-analysis. (Harman 2007, 192)

Harman's emphasis on what he calls sensual and real objects led him to christen his approach "object-oriented ontology," borrowing the phrase half-jokingly from the object-oriented approach to programming in the world of computers. The name stuck and OOO has begun to have significant cultural cachet beyond the bounds of academic philosophy, especially in the areas of architecture and art.

In his 2018 book *Object-Oriented Ontology: A New Theory of Everything*, Harman sums up his "basic principles" as follows:

> (1) All objects must be given equal attention, whether they be human, non-human, natural, cultural, real or fictional. (2) Objects are not identical with their properties, but have a tense relationship with those properties, and this very tension is responsible for all of the change that occurs in the world. (3) Objects come in just two kinds: real objects exist whether or not they currently affect anything else, while sensual objects exist only in relation to some real object. (4) Real objects cannot relate to one another directly, but only indirectly, by means of a sensual object. (5) The properties of objects also come in just two kinds: again, real and sensual. (6) These two kinds of objects and two kinds of qualities lead to four basic permutations, which OOO treats as the root of time and space, as well as two

> closely related terms known as essence and eidos. (7) Finally, OOO holds that philosophy generally has a closer relationship with aesthetics than with mathematics or natural science. (Harman 2018, 9)

We will unpack these further as we proceed.

Points of Agreement

There are a number of uncontroversial points of agreement between Harman's OOO and the writings of Polanyi himself along with the dominant interpretations thereof. They are, first and foremost, both enemies of reductionisms and materialisms of all sorts. "For the object-oriented thinker," says Harman, "physical objects are just one kind of object among many others…hence we should not be in a hurry to scorn or 'eliminate' those that are not a good fit with a hardnosed materialist worldview" (2018, 39). Thus "philosophy is not the handmaid of materialism any more than of religion" (Ibid.). He is thus "also of the opinion that materialism must be destroyed," (2010).

Any attempt to ontologically dismiss an entity by reducing it to constituent parts is anathema for OOO and such reductionist moves are often referred to as "undermining." At the same time, OOO guards carefully against the opposite philosophical error, "overmining," which is the move in which middle-sized entities are dissolved into an undifferentiated, universal whole (such as Heidegger's World) – "the only necessary criterion for an object in OOO is that it be irreducible in both directions: an object is more than its pieces and less than its effects (Ibid., 53). Harman also points out that over- and undermining can be combined in complex ways, which he calls "duomining." The idea is to ontologically accept a full panoply of objects, including fictions, and to maintain such objects' distinctiveness while also affirming their vicarious relational interconnectedness at the level of sensual objects, an aim which is compatible in significant ways with Polanyi's conception of comprehensive entities, a topic we will explore further later on.

OOO also takes a moderate emergentist position, affirming that an "emergent reality of an object composed jointly of multiple parts (such as a married couple) does not hinge on the predictability or unpredictability of how it ultimately turns out…[since] emergence does not require mysterious results, but only that the married couple has joint features not found in either of the individuals in isolation" (Harman 2018, 31). Given the emphasis on emergence theory by Polanyi and many of his interpreters, this is another important point in common.

Finally, there is sort of affirmation of the spirit of the IFM test of reality that permeates Harman's work, though it is never explicitly theorized. The possibility of indefinite future manifestations is concomitant with and implied by Harman's theory of the real object, which lies beyond our reach, but which interacts with us through the sensual object that we do directly access. We note the resonance too when Harman digs into the tool-analysis of Heidegger, one of his primary forerunners, noting that "we are no more obliged to follow Heidegger's own understanding of tool-being than we are to follow the itinerary of Lewis and Clark when driving from St. Louis to Portland, or to limit our uses of electrical power to those devices patented by Thomas Edison himself" (Harman 2002, 16).

There are clear echoes here of the sentiments expressed by Polanyi throughout his corpus, found, for example, wherever he traces the ongoing development and refinement of Copernicus' insight by Galileo, Kepler, Newton, and beyond. Namely, the sentiment so expressed is that "the historical greatness of explorers or inventors or philosophers does not guarantee that they have exhausted their own subject matter" (Ibid.). The Polanyian gloss on the matter goes further and notes as well that the historical greatness of such figures is proportional to the degree to which they were unable to exhaust their discoveries due to the subject matter's profundity.

Heideggerian Roots

As referenced earlier, OOO spins its distinctive ontology out from the intentional object of Husserl, giving rise to the sensual object, and the tool-analysis of Heidegger, leading to the theorization of the real object. The sensual object, for the time being, can be taken at face value – it is the object of intention that is directly available and susceptible to phenomenological analysis. At first glance this may lead us to take the real object as simply a new name for Kant's thing-in-itself, the noumena eternally cut off from the phenomenal realm. There is an important kernel of truth to this, but such a reading would nevertheless be overly hasty. To see how the picture OOO paints is different from that of the Kantian divide, we will briefly examine its roots in Harman's reading of Heidegger's tool-analysis.

For Harman, the most important facet of Heidegger's philosophy is *Zuhandenheit* or readiness-to-hand, which he rechristens *tool-being*. Yet Harman argues that Heidegger's analysis of equipment applies to all entities, not just to tools traditionally conceived. He notes that equipment is not effective "'because people use it;' on the contrary, it can only be used because it is capable…of inflicting some kind of blow on reality…in short, the tool isn't 'used' – it is" (Harman 2002, 20). Equipment thus forms a background or landscape that structures reality through the constraints and opportunities it affords. In Polanyian terms, equipment sets up a gradient that shapes possibility wherein "beings in themselves are ready-to-hand, not in the derivative sense of 'manipulable,' but in the primary sense of 'in action'" (Ibid.).

The presence-to-hand of beings (that is to say, their status as held within a theoretical gaze) is more or less mappable onto the sensual objects derived from Husserl's intentional object, while the tool-being, or readiness-to-hand is indicative of the real objects, punching up from the depths to make their moves.

On Harman's reading, then, all being (in the sense of real objects) is equipment, and all equipment has the following two facets:

> (1) its irreducibly veiled activity, and (2) its sensible and explorable profile. In more familiar Heideggerian terms, there is the tool viewed "ontologically" and the same tool viewed "ontically." For the moment, we have no way of bringing these worlds into communion, other than to say that one is primary and the other not primary. (Ibid., 22)

The theoretical work of OOO is largely defined by the task of speculating on how these worlds do commune, as well as by arguing that contrary to the common assumption amongst thinkers since Kant, "the true chasm in ontology lies not between humans and the world, but between objects and relations…[and that] this duality holds equally true for all entities in the cosmos, whether natural, artificial, organic, or fully human" (Ibid., 2). Thus entities are contacted by knowers or by one another in one sense, but withdraw from presence in another sense, distancing "themselves not only from human beings, but from each other as well… forever haunted by some hidden surplus…that never becomes present" (Ibid.). Yet this hidden surplus is not necessarily static, and so, to put a Polanyian spin on things, the surplus can go on becoming present in starts and fits, giving rise to the IFMs central to Polanyi's thinking.

Summing up OOO's basis in tool-being, Harman recurs to one of his often striking metaphors:

> It is the nature of tool-being to recede from every view…Like the giant squids of the Marianas Trench, tool-beings are encountered only once they have washed up dead on the shore, no longer immersed in their withdrawn reality. It is impossible to define tool-being as a linguistic network or culturally coded system of "social practices,"…Tool-being is that which withdraws from all such networks…Hence, what Heidegger opposes to theory is not human praxis, but a mysterious capital "X," a brutal subterranean realm which we can glimpse only at second hand…But by the same stroke, he unknowingly suggests a possible campaign of guerilla metaphysics. Tool-beings turn out to be a strange variant of traditional substances, though they are as irreducible to physical particles as they are to the traces they leave in human perception. *They are substances that exceed every relation into which they might enter, without being ultimate pieces of tiny matter. But this leaves only one possibility: for the first time in a long while, Heidegger pushes philosophy to the point where it has no choice but to offer a renewed theory of substantial forms.* (Ibid. emphasis added)

It will be my contention that there is some significant overlap between OOO's project of proposing a renewed theory of substantial forms that exceed every relation and are irreducible to material parts and Polanyi's sketches of an ontology of comprehensive entities that stake their metaphysical claim in Eleatic terms, through their IFM production.

Ontology: The Quadruple Object and "Contact" with Reality

Tool-Being, on which I have been drawing in describing OOO's basis in Heidegger's analysis of equipment, is Harman's first book. In subsequent works he has blended and developed these Heideggerian themes in ways which I will attempt now to briefly sketch, before jumping in more depth into selected areas of overlap and tension with Polanyi's thought.

On OOO's account the real object instantiates tool-being as explained above, receding from view, yet existing for all that with its real qualities. The sensual object with its sensual qualities is what we experience directly, the object of our intention. What tissue connects the two? This remains mysterious. The only real object we can access directly is that of our own person. We cannot approach other real objects directly, but we do seem to have intimations that they are out there, in the Marianas Trench, as it were. We must then make our approach to connecting the sensual to the real, *aesthetically*, which is to say indirectly, and to some extent unaccountably, since there is an abiding tension and paradox in OOO's twin aims of "preserving the inaccessibility of things-in-themselves while also defending the claim of art to make contact with the executant inwardness of these things" (Harman 2018, 52). We are, he claims, "touching, without touching" (Ibid.) Together, these four elements make up a system that Harman calls the "quadruple object":

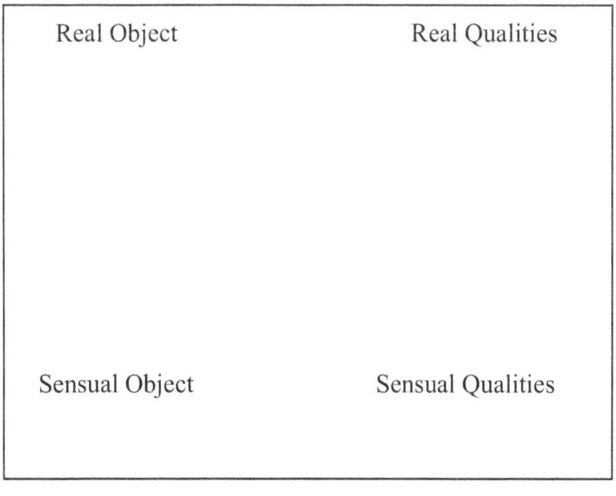

The quadruple object

What is most specially of interest to Harman and adherents of OOO are the "bonds or tensions" (Harman 2018, 159) between these four poles, rather than the poles themselves. These tensions give rise to a set of four analytic terms – space, time, eidos, and essence – which I shall briefly enumerate and explain with recourse to Harman's own descriptions.

Space names the tension between the real object and the sensual qualities united to it by means of the quadruple object. Harman explains that

> the reason for this is that space means both proximity and distance, just as is true of RO – SQ. Whatever is in space is removed from us and ensconced in its own private place, but also belongs to the same spatial arena as we do…[its] determinate distance from us…might be overcome if sufficient energy were expended. (Ibid.)

Time tracks the tension between the sensual object and its sensual qualities, which are constantly on the move from one instant to the next (for instance as we move around a table, looking at it from different angles, etc.) Harman, in discussing this tension which is so central to Husserl's work, notes that

> Just as space requires both proximity and distance, time entails both endurance and change: if we encountered nothing but a strobe – like kaleidoscope of constantly changing properties, this would be an experience of derangement rather than of time. What time gives us, instead, is the experience of constant flickering change amidst a more slowly shifting background of enduring sensual objects. (Ibid., 161)

Eidos, echoing Plato through the prism of Husserl's work, names the tension between the poles of the sensual object and real qualities. This tension.

> concerns Husserl's insight that sensual objects do not just have shifting sensual qualities, but indispensable real qualities as well…Here we encounter the strange fact that a sensual object (which exists only as a correlate of our paying attention to it) nonetheless has real qualities (which exist whether we are aware of them or not). (Ibid.)

Essence is the most elusive of these four terms, naming, as it does, the tension between the real object and its real qualities. OOO rejects postmodern arguments to the effect that essences are oppressive, since it holds that this can only be the case if one claims to have grasped the essence of a thing. Thus Harman:

> It should never be forgotten that OOO renounces all claims to know the essence of anything directly. Yet this does not entail that nothing has an inner nature, and that therefore everything would be willfully performed or socially constructed rather than having an inherent character that resists all our efforts to plumb its depths. Since essence is the only one of the four tensions made up of two real terms and no sensual ones, it is admittedly…the most difficult…to investigate. (Ibid., 162)

Now, the question remains of how, on OOO's account, we penetrate, albeit indirectly, the subterranean depths to make some sort of contact with real objects and real qualities. Harman's answers to this question are strange and fascinating, lending credence to his tendency to refer to his own position as a "weird realism" (2012) We dive now into a brief exposition of Harman's proposed solution to this Kantian conundrum.

We will make our approach by way of Harman's understanding of metaphor, which is central to his thought, and which he opposes starkly to literal language. He is fond of using Homer's suggestive phrase, "the wine-dark sea," as an example upon which to work, identifying the sea as a sensual object, which is being tied metaphorically to the sensual quality of the darkness of the wine. From there let us see how he navigates this interplay between the beholder and the two quadruple objects (wine and sea) at play in the metaphor:

> …it is impossible for us to ascribe wine-qualities literally to the sea, or we would simply have a literal statement… As a result, though the wine-qualities remain largely familiar as they have always been, the sea becomes a mysterious sea, capable of holding wine-qualities in its orbit. Another way of putting it is that SO (sea) is replaced by RO (sea). (Ibid. 182)

If this last move is confusing, remember that on OOO's account the real object is shrouded in mystery. Thus when the conjunction of wine-qualities with the sea causes the sea to cease to be a large body of saltwater, easily defined in literal terms, it throws us towards the mysterious real sea. Harman's analysis continues,

> Thus, the metaphor should work to combine RO (sea) and SQ (wine-qualities). But now we come to the final stage. RO (sea) cannot possibly participate in the metaphor, since every RO withdraws or is withheld from any access or relation whatsoever. This, in turn, should have the implication that SQ (wine-qualities) is floating in space, attached to nothing. But the most important lesson we learned from phenomenology is that there is no bundle of qualities unattached to an object. And since RO (sea) is permanently unavailable, it is replaced by RO (beholder). I myself, as the reader of the poem, must perform the metaphor by standing in for the absent sea, or else no metaphor results. Rather than burrowing downward towards the reality of the sea, the metaphor builds a new theatrical sea-reality upward, on top of the sensual qualities of the wine. There is no 'justification' for believing in a wine-dark sea, since it is not built from a literal comparison. And yet this belief cannot possibly be untrue, because it creates its own object in the very act of believing it. This is why it makes sense to call aesthetics a case of 'unjustified true belief,' though one that is not a form of knowledge. (Ibid., 182)

Many Polanyians, myself included, will likely balk at the stark division between literal and figurative, aesthetics and knowledge, but let us table our concerns for the moment. We will address these issues below. For now, let us see how Harman's take on metaphor is instructive in guiding us toward an understanding of how OOO proposes that a sort of contact with underlying realities is made. In the following passage Harman considers a scenario in which a beholder is looking on the sun (the sensual object with its sensual qualities), attempting to dive down, so to speak, to the real sun with its real qualities.

> We therefore must ask: what RQ is already available on the scene in the same way that RO (beholder) was when it stepped in and saved the day for metaphor? The RQ (sun-qualities) cannot be directly known, since…they withdraw every bit as much as does RO (sun). Nor can the needed RQ be

> found in any of the other things surrounding the sun in experience, since these too are sensual objects manifesting nothing but sensual qualities. Once again the beholder must do the job, yielding a combination of SO (sun) and RQ (beholder-qualities). To repeat, this is not theatrical work by the beholder in the metaphorical manner in which RO (beholder) is combined with SQ (wine-qualities), since we are not ourselves stepping in as objects the way we did when replacing the absent sea of Homer. Instead, it is somehow the case that real qualities drawn from us are lent to SO (sun)...This is why knowledge does not require the same degree of personal absorption as aesthetic experience: we can look on dispassionately at a riddle of knowledge rather than constantly sustaining it with our own being as we must in aesthetics. As soon as we are bored the artwork is no longer an artwork, but boredom in matters of knowledge is not a rare case at all, and by no means does it destroy the object of knowledge. (Ibid., 185).

To sum up, an OOO account holds that the "contact" (or, perhaps, the beholding) that takes place when we chase a metaphor upwards or burrow down toward a subterranean real object comes about by a family of substitutionary moves by the beholder. This has consequences for concepts like truth and knowledge, since "if truth means a correspondence between our beliefs and reality, or some subtler variant thereof (such as Heidegger's partial 'unveiling'), then OOO...cannot possibly accept truth as a criterion of knowledge" (Ibid., 181). Furthermore,

> knowledge (unlike aesthetics) concerns a sensual object rather than a real one, but also aims to arrive at the real qualities of its sensual object rather than any sensual qualities. We also said that knowledge cannot be metaphorical but must be literal, in the sense that the object of knowledge – such as SO (sun) or any other SO – must be reducible to its qualities in such a way that it amounts to nothing but a bundle of components or effects. (Ibid., 186)

Science, as a knowledge-seeking enterprise (in OOO's terms) works primarily on a horizontal plane and deals in justified untrue belief (since it cannot establish in literal, explicit terms any link to underlying realities). Meanwhile, aesthetics makes vertical moves, both up – via metaphor, and down, vicariously, towards underlying realities. This is the arena of unjustified true belief.

Constructive Tensions: Epistemology and Metaphysics

There is much here to take issue with, from a Polanyian standpoint, however it is important to note that OOO is grappling with many of the same issues that Polanyi attacks in his own work, ultimately, with similar aims. For OOO, it must be recalled, is not proposing antirealism, but is rather seeking to reform and expand our understanding of reality by proposing a new theory of substances. Its approach is by way of metaphysics, taking much of the problematic language of critical epistemology for granted and then making surface-level terminological changes when its developing metaphysics exposes a problem within the critical epistemological paradigm. Polanyi is likewise concerned to put his metaphysical conception of Copernican realism back on the table as a live option. He makes his approach, however, by attacking the canons of critical epistemology, and setting up his theory of tacit knowing in its place, a project that begins with a radical shift in the definition of knowledge itself. The move to metaphysics for Polanyi comes, explicitly and terminologically at least, as a result of his epistemological work, though it motivates his entrance into epistemology from the start, as I will soon argue. So OOO moves with realist intentions from being to knowing while Polanyi slides from knowing to being, with similar motives.

In what follows, I shall attempt to examine some of the constructive tensions between these two approaches, indicating where the two movements might profit from the others' insights. In short, I will argue that OOO is in need of an epistemological shot in the arm from Polanyi, while OOO can provide some important resources for a more thoroughgoing development of a metaphysics of comprehensive entities than Polanyi provided.

In an epistemological sense, OOO attempts to overcome the strict Kantian divide between the noumenal and phenomenal while maintaining an appropriate humility with regard to our knowledge of underlying realities *and* respecting and celebrating our creative participation in related noetic processes, granting to these some sort of purchase on reality. Again, this is an attack on reductionisms of all sorts, and for Polanyians, constitutes a very respectable set of aims. The twofold question that arises is whether OOO's insistence on the stark division between knowledge as literal and explicit (and by implication uncreative) on the one hand, and aesthetics as metaphorical, implicit, and creative, on the other, are (1) theoretically defensible in light of experience, and (2) likely to be rhetorically effective in terms of clarity and perspicuity.

With regard to # 1 above, a Polanyian critique will lead to the conclusion that OOO's strict epistemological divide does not hold up to scrutiny, but rather substitutes a binary in place of a more complicated continuum. The critique begins with Polanyi's insistence on the tacit (and hence inexplicit) dimension of all knowledge. Even the dullest of scientific propositions – "The Earth is round," – taps into a vast and only partially articulable and explorable network of subsidiary details. It is trite and unexciting to *us, now* in its literal obviousness, but to Columbus, setting sail for the Indies, it seemed far from settled. Yet he was willing to hazard life and fleet on its truth. In Polanyian terms then (while borrowing some of the poetic verve of Harman's prose), there is a rich, tacit underworld lurking just on the other side of the driest of facts, propositionally expressed. This angle of critique can be deepened and thickened by referencing Polanyi's work on the heuristics of discovery, as the reference to Columbus indicates, for such facts and propositions do not exist in a vacuum, but are themselves usually the record of some process of discovery, whether this be the result of a laboratory experiment or the dawning realization of a baby that "doll" means *doll*. Furthermore, Polanyi has shown persuasively that Eros is the gocart of discovery that follows the track laid down by beauty.[11] Thus Polanyians will generally insist on the centrality of the aesthetic dimension for all noetic pursuits, including those, such as science, that fall on the "knowledge" side of the OOO binary.

Thus, to deliver Polanyi's point in OOO's terms, there is no "knowledge," only aesthetics, and this brings us to question # 2, on the clarity and perspicuity of OOO's terminology. The claim that there is no knowledge, but only aesthetics, simply cries out for an antirealist gloss, which,

[11] "The affirmation of a scientific theory was seen to convey an appreciation of its beauty and all the statements of mathematics were seen to carry a whole gamut of delicate aesthetic appreciations" (PK 204).

as we have seen, is the reverse of what OOO is driving at. This is in part due to the history of Western thought, but it is also due in large part to our use of the verb "to know" in ordinary language to cover everything ranging from perception to skills to propositions, and beyond. Arguably, then, Polanyi's reformation of epistemological terms leads to clearer, more perspicuous epistemological analysis, since instead of forcing everyone to rigidly demarcate the bounds of "knowledge" and "aesthetics" in speech and thought (an unlikely scenario, to say the least), it allows for consistent, everyday usage at the level of ordinary language while gradually reforming the metaphysical underpinnings of that language in an ongoing, largely tacit, process (this is perhaps improbable as well, but it is in my view a likelier scenario given that it conforms more closely to the way language and concepts gradually evolve).

We have not yet commented on the personal coefficient in Polanyi's epistemology as it might impinge on the OOO approach. My contention here is that we find significant overlap, for, if we recall Harman's explanation of the beholder's substitutionary moves in the act of metaphor creation and in attempting to connect somehow with real objects and qualities below the decks of the sensible world, what we see is the enactment of a participatory realism, perhaps quite consonant with Phil Mullin's reading of Polanyi's partial metaphysics. Part and parcel of the personal coefficient of knowing are commitment and indwelling, as is well known to those versed in Polanyi's thought, and throughout Harman's work there are echoes of both of these, given the centrality, for OOO, of the "vanishing object that needs replacement by *an engaged beholder who takes on the duties of the missing object*" (Harman 2002, 107 emphasis added).

Elsewhere, building on Levinas's insight that the world is a web of nourishments, Harman argues.

> Consciousness does not penetrate back behind the objects that surround it; it does not outflank the countless elements that bombard it like snowflakes or hailstones. Instead, consciousness is sincere. Life is fundamentally an enjoyment. It is nothing but an absorption or involvement…(Harman 2002, 241)

"Enjoyment" and "absorption" point to our inescapable integrative capacity for indwelling, while "sincerity" highlights the authenticity of our commitment, even as "involvement" gestures simultaneously toward both.

Yet there is an important sense in which OOO's conceptual framework could benefit from its own absorption of a more thoroughly Polanyian take on the committed and sincere personal participation of the knower/beholder, for Polanyi is able to make some headway toward resolving the tension between the poles of the objective/subjective dichotomy by the introduction of personal commitment with universal intent, while OOO more or less accepts the language of the dichotomy while pushing back against its dominant modernist interpretations in various ways. While OOO's pushback on this front is often cogent and effective, it still leaves us, at the end of the day, with a conceptual apparatus that, at the linguistic level at least, drives us toward a Cartesio-Kantian bind: Knowledge is objective, and aesthetics are subjective, yet for OOO knowledge is confined to the horizontal plane of sensual objects and sensual qualities – a plane which is itself subjective. The committed OOO theorist will not interpret it this way, and there is a sense in which they are right not to, since their project is, like Polanyi's, an anti-scientistic and non-relativist one. Nonetheless, I see this as a place where OOO might benefit from a more detailed epistemological development, and Polanyi's analysis and complication of the objective/subjective dichotomy has much to offer here.

Now, it may well be that adherents of OOO would see Polanyi's emphasis on the personal coefficient as dangerously anthropocentric. Polanyians, meanwhile, many self-identifying as personalists, would sniff suspiciously at OOO's mission to take seriously at a metaphysical level interactions between non-human objects, but I think this would be too hasty in both cases. In part this is, again, because Polanyi's systematic work is first and foremost an epistemology, and Harman's a metaphysics. Polanyi's emphasis on the person is thus primarily epistemological, though it is also, to a lesser degree, metaphysical (as seen in the fourth part of PK in which humanity is presented as the highest development yet of the universe known to us).

This epistemological and metaphysical pride of place given the person in Polanyi's work may well be objectionable to the OOO theorist, but I do not think it should preclude productive discussion between the two camps, especially since there remains in OOO, it ought to be pointed out, an inescapable anthropocentric epistemological bias insofar as it is *a metaphysical theory generated by people reflecting on the world.* This, however, in no way undermines its cogency nor its interest in taking account of non-human swathes of the world in a more attentive way than much of the continental tradition has done. Thus in my view, potential accusations of anthropocentrism and of objectivist depersonalization alike ought to be tabled for the time being.

It is time, however, to wrap up this brief epistemological critique of OOO and to look at some possible ways in which OOO's metaphysics might help to flesh out Polanyi's notably unsystematic metaphysics. It is thus to this area that we now turn.

For Polanyi, it will be recalled, the from/to structure of tacit integration is the mechanism whereby a set of subsidiary particulars points toward its joint meaning, a comprehensive entity, taken to be real to the extent that it is capable of producing IFMs. Yet there is a two-way directionality to the from/to structure, since the parts influence the whole, but the whole also influences the parts, and our attention can be run back and forth between the two poles. As we will soon examine in more depth, Polanyi thus makes the jump to metaphysics by means of the ontological aspect of tacit knowing, whereby he affirms that the metaphysical structure of reality corresponds to the subsidiary particular/comprehensive entity relationship, a theory he develops in terms of emergence, boundary conditions, etc.

While Polanyi makes various metaphysical claims throughout his work, he is only relatively systematic when it comes to his epistemology. This is, as already noted, the mirror image of OOO's approach. One aim of this work is to develop a more complete metaphysics that harmonizes post-critically with Polanyi's epistemology, beginning with an ontology of comprehensive entities. As will be seen in Chap. 3, Harman's analysis of tool-being and the quadruple object will be of some service in this task, given their similarity in important respects to Polanyi's conception of comprehensive entities.

Polanyi, it will be remembered, wanted, by an epistemological route, to arrive at a "resurrected idea of reality" which would "look different from its departed ancestor" –

> Instead of being the clear and firm ground underlying all appearances, it will turn out to be known only vaguely, with an unlimited range of unspecifiable expectations attached to it. (1967, 177)

OOO, led by Graham Harman has a similar program – to rehabilitate the concept of substance as the basis of a non-modern metaphysics (following here Bruno Latour's usage). My aim here, as with the previous section on Ontic Structural Realism, has been to put Polanyi's work into conversation with a range of realisms that continue to serve as cruxes of contemporary metaphysical debates, both continental and analytic. In the case of OOO, I would argue that Polanyians might find an important sparring partner and erstwhile ally for the post-critical project.

3.3 Neo-Aristotelianism and Contemporary Hylomorphism

We return now to the world of analytic philosophy, where for a number of years a resurgence in interest in Aristotle has led to a rich output of new metaphysical material by thinkers such as Kit Fine (1999), Mark Johnston (2006), Kathryn Koslicki (2008), David Oderberg (2005), Michael Rea (2011), Anna Marmodoro (2013), Robert Koons (2014), William Jaworski (2016), and William M.R. Simpson (2017, 2023). Much of this work has significant overlap with Polanyi's own philosophical project, though it would, in my view, be incorrect to categorize Polanyi as a neo-Aristotelian, primarily due to his position on universals, as will be seen.

Despite the existence of important internal disagreements,[12] neo-Aristotelian thought offers a thoughtful, non-reductionist metaphysical realism which centers around contemporary interpretations and develop-

[12] See Chap. 2 of Simpson (2023) and Johnston (2006), pages 659–660 for a summation of key disagreements within the hylomorphist camp.

ments of Aristotle's doctrine of hylomorphism, which holds roughly that substances are compounds involving form and matter. For example, in *De Anima* Aristotle considers what sort of analysis would be most appropriate to the physicist, in the process giving a succinct statement of basic hylomorphic theory:

> [One type of physicist] assigns the material conditions, the other the form or account; for what he states is the account of the fact, though for its actual existence there must be embodiment of it in a material such as is described by the other. Thus the essence of a house is assigned in such an account as 'a shelter against destruction by wind, rain, and heat'; the physicist [5] would describe it as 'stones, bricks, and timbers'; but there is a third possible description which would say that it was that form in that material with that purpose or end. Which, then, among these is entitled to be regarded as the genuine physicist? The one who confines himself to the material, or the one who restricts himself to the account alone? Is it not rather the one who combines both? (403b1–5)

Contemporary hylomorphists argue powerfully that Aristotle is right on this score, urging that substantial form or structure is an irreducible and fundamental aspect of material reality that must be taken account of. In the introduction to *Neo-Aristotelian Perspectives on Contemporary Science*, William M.R. Simpson, Robert Koons and Nicholas J. Teh offer the following description of the generally agreed-upon tenets of neo-Aristotelian thought:

> [1] First, neo-Aristotelian philosophers count the concept of potentiality as an essential feature of their metaphysics…embrac[ing]…a causal powers ontology, in which both active and passive powers are regarded as fundamental features of particular things in the world that bring about change…[This] entails the reality of some type of teleology, since powers…[are] directed toward possibly unrealized (future) actualities.[2] Secondly, a neo-Aristotelian account must include what Jonathan Schaffer has called a 'layered' or 'structured' view of reality, insofar as some entities and properties are regarded as being more fundamental…and other[s]…to be derived…The fundamental entities, according to some neo-Aristotelians, are basic *substances*; according to others, certain primary *powers*.

> [3] Thirdly, a neo-Aristotelian account of reality is not monistic but involves a plurality of entities, both simple and composite. The whole cosmos is not the only substance, nor is everything made from a single set of simple substances, but the world also includes natural unities like biological organisms, which are composite substances.
>
> [4] Fourthly, substances in nature belong to recurring natural kinds, each with its own intelligible nature or essence. These commonalties are not subjective, conventional, or wholly mind-dependent…
>
> [5] Finally, neo-Aristotelians reject extreme realism or Platonism.[13] Neo-Aristotelian accounts of nature include no appeal to non-immanent, non-natural universals; they instead hold that mathematical models of the physical world should be regarded as idealizations that invariably involve an empirical loss of scientifically relevant information. (Simpson et al. 2017, 1–2)

I will now proceed to briefly consider Polanyi's own thought in terms of the five points of neo-Aristotelian agreement outlined above. This will require me to anticipate to some extent the interpretive and

[13] At this point I must introduce a somewhat unfortunate terminological binary that will be necessary for clarity off and on throughout the remainder of this book. The terms I refer to here are "*Anti-Platonist* Aristotelianism" (including references to the Aristotle of Anti-Platonist Aristotelianism) vs. a *Platonist* Aristotle (which gestures toward a harmonization of Plato and Aristotle similar to the Neoplatonist synthesis and which therefore, in my view, is a subspecies of Platonism). The need for this terminological hack arises due to the fact that the most prominent expositions of Aristotelianism generally take Aristotle to be in conflict with Plato, especially with respect to the question of universals and Forms, as seen in point 5, cited in the main text above. The Anti-Platonist Aristotelian position, glossed in more depth later, holds that there are no transcendent, uninstantiated universals (i.e. Platonic Forms) but is realist with respect to immanent universals which are instantiated in particular sensibles. Such an Aristotelianism will also often show a sort of ontological favoritism toward particulars (primary substance) over universals (secondary substance) which is based on Aristotle's *Categories* (see also Norio Fujisawa 1974, 38–40 on such an "Aristotelian thought framework").

There is, of course, plenty of variety in viewpoint among those who consider themselves Aristotelians, however this is, in my estimation, a more or less fair summation of what is taken to be the Aristotelian position on universals in most mainstream debates, philosophy classes, etc. When I critique Aristotelian positions in what follows, I am generally critiquing this Anti-Platonist Aristotelianism. Interspersed throughout the book, however, (and coming more into focus in the final chapter) are references to a Platonist reading of Aristotle which takes Aristotle himself to be a Platonist interested in developing a more rigorous understanding of matter in harmony with the central doctrines of Platonism (see Gerson 2005). On this reading, his theory of universals is taken to be an explanation of our conceptual tracking mechanism and *not* a rival metaphysical theory to the Forms.

developmental work on Polanyi's metaphysics that I will lay out in detail in subsequent chapters. I justify this somewhat serpentine strategy by noting that insights drawn from contemporary hylomorphist thought will be woven into those remaining chapters as appropriate.

Point 1: What would Polanyi have to say about the neo-Aristotelian emphasis on potentiality as bound up with the active and passive powers that are taken to be fundamental aspects of natural entities, and the teleology that such powers entail, given their directedness?

He would certainly be in significant agreement with contemporary hylomorphists on these matters. Potentiality figures prominently in Polanyi's thought, both at the level of the IFMs that establish the reality of entities in accord with the Eleatic Principle (in essence another terminology for active and passive powers) and in terms of his field theory, which postulates entities (fields) that consist of "the gradient of a potentiality: a gradient arising from the proximity of a possible achievement" (PK 398) which cumulatively amount to a "gradient of deepening meaning" (*M* 177). The potential achievement is here taken to be a real entity given that the gradient it produces is an IFM, an expression of its Eleatic standing.

Polanyi also explicitly affirms the teleological nature of comprehensive entities. One instance of this can be seen in his teleological approach to biology facilitated by his conception of living beings as *centers of achievement* (*PK* 349, 354, 357, 400, 404). Potentiality and telos as bound up with Polanyi's fields and his understanding of comprehensive entities will serve as centerpieces of the following chapters.

Point 2: Does Polanyi affirm a layered view of reality which holds some entities and/or properties to be more fundamental than others, with certain other entities or properties emerging therefrom?

The simple answer here is that yes, Polanyi does affirm such a view, but as is usually the case, the devil is in the details. Polanyi is an explicit emergence theorist in his fight against the reductionist tendencies in modernity, and insofar as we focus on his theory of boundary conditions and dual control, he seems to theorize a stratified universe along these lines.

His theory of comprehensive entities, however, lacks the Aristotelian stipulations that all substances are (a) particulars and (b) natural kinds (i.e. excluding artefacts, etc.). Since the reality of a comprehensive entity is primarily established for Polanyi by its ability to effect IFMs, some of which result in the emergence of new comprehensive entities, this stratification begins to be problematized. Polanyi himself does not provide a systematic working-out of this conflict and there are therefore conflicting interpretations of his position here, as we have already seen. For example, Prosch and Gulick urge an ontological distinction between the real, which is founded on material substrates, and meaning, which is emergent but not ontic, while other interpreters oppose this vigorously, arguing that Polanyi gives equal ontological weight to *all real entities*.

Nonetheless, even with the caveats sketched above (to be developed more completely anon), I think it fair to say that within the realm of concrete, sensible particulars Polanyi is in general agreement insofar as he affirms real, emergent entities that arise within the context of prior entities which might be construed as more fundamental in certain senses. Thus for Polanyi a living thing that devises an artefact is a more fundamental entity than the artefact, but the artefact is still a substance. Polanyi would likely argue that the living thing is *more real* or a *richer* substantive entity due to its greater potential for IFM production, but his account would lack the rigid lines of substantival demarcation characteristic of much Aristotelian thought.

Point 3: Would Polanyi agree with neo-Aristotelians that reality is not monistic but rather includes a variety of entities, including composite substances?

Polanyi would, I think, readily agree here. His conception of the universe clearly includes a multiplicity of real entities capable of IFM production. Nonetheless he would be likely, on my reading, to allow for more types of substantive entities than many contemporary hylomorphists, for the reasons glossed under point 2 above. Furthermore, given the close connection between knowing and being and our ability to oscillate between focal and tacit/subsidiary attention as affirmed by Polanyi, the boundaries between substantive entities might be more amorphous than in neo-Aristotelian thought.

Point 4: Would Polanyi agree that "substances in nature belong to recurring natural kinds, each with its own intelligible nature or essence" and that these similarities are not merely "subjective, conventional, or wholly mind-dependent" (Simpson et al. 2017, 1–2)?

He would wholeheartedly concur on all of these points. Furthermore, in pushing back with neo-Aristotelians against anti-realism whether it comes in the guise of objectivism, subjective idealism, or overly conventionalist forms of pragmatism, Polanyi makes constructive and distinctively post-critical moves toward problematizing the subjective/objective dichotomy by introducing the personal coefficient and the concept of universal commitment. Additionally, he would appreciate the inclusion of the adverb *wholly* above, since his theories emphasize the participation of the knower with the known, including – in some cases at least – at the constitutive level.

Point 5: Finally, would Polanyi reject Platonism with regard to universals, often referred to as extreme realism, preferring, with neo-Aristotelians, to countenance the existence only of instantiated, immanent universals?[14]

Polanyi is undoubtedly a realist with respect to universals, as are most Aristotelians. The question is whether he restricts this realism to instantiated, immanent universals or affirms, in contrast, the reality of uninstantiated universals. Many interpreters of Polanyi's corpus, thinking of his statement that universals "are the joint meaning of things forming a class" (KB 170) which are approached inductively, would say "yes," thinking of the *prima facie* similarity of this statement to the Aristotle's discussion of universals in *Posterior Analytics.*

In a subsequent chapter on Polanyi's field theory and its implications for his theory of universals, however, I shall argue that he is, in fact a Platonist[15] with respect to universals, affirming both implicitly and explicitly the existence of uninstantiated universals.[16] I will not rehearse my arguments for this conclusion here, since we will come to them soon

[14] While most neo-Aristotelians are realists with respect to immanent, Aristotelian universals, there are exceptions. William Jaworsky (2016), for example, is a trope theorist, and thinkers such as Koons and Simpson are moving in that direction as well (email message from W.M.R. Simpson, May 7, 2023).

[15] Charles Lowney has also noted Polanyi's Platonist stance on universals, especially in "The Tacit in Frege" (2008), and holds that there need be no real conflict between Aristotle and Plato on this point. I agree with him on both counts, as will be made clear in subsequent chapters.

[16] The common understanding of "universal" either as a rough synonym for "Form" (or as a rival term for describing the metaphysical ground of entities' natures) will be called into question in subsequent chapters. For now, however, we will stick to the standard terminology as it appears in neo-Aristotelian scholarship.

enough. Nonetheless, I will note in passing that the Platonism inherent in Polanyi's overall project is *not* the type of Platonism commonly discussed in analytic circles insofar as it *does not* posit universals as otherworldly, purely mathematical entities which necessarily stand outside causal relations.

On balance, then, Polanyi is in basic agreement with neo-Aristotelian thought on points 1, 3, and 4 above. On point 2, I would say he is in agreement with some important caveats, while on my reading he would reject the neo-Aristotelian position on universals covered under point 5 – though as I have already mentioned, other interpreters of Polanyi may well disagree with me here. The arguments to be taken up soon regarding Polanyi's position on universals are important, however, since if I am correct that he affirms uninstantiated universals, he is some type of Platonist, whereas if my arguments fail, it would be fair to identify him only as a somewhat idiosyncratic fellow traveler in the neo-Aristotelian camp.

Either way, the overlap between neo-Aristotelian realism and Polanyi's Copernican realism is striking and thus signals the presence of an important philosophical ally and source of insight for a post-critical Platonism growing out of Polanyi's project. In the interest of fostering such a constructive dialogue, I now propose to briefly examine some similarities between Polanyi's conception of the comprehensive entity and neo-Aristotelian notions of hylomorphic compounds before looking in additional detail at contemporary interpretations of the Aristotelian understanding of substantial form, which will provide a helpful framework for working out a more robust metaphysics of comprehensive entities in the following chapter. It must be noted that neither of these efforts will be exhaustive, but are rather intended to provisionally sketch the details of some constructive points of contact between Polanyian and neo-Aristotelian metaphysics.

Comprehensive Entities, Hylomorphic Compounds, Joint Meanings and Substantial Forms

Polanyi's conception of the comprehensive entity (or joint meaning) that we contact by means of the tacit integration of the relevant subsidiary particulars has certain obvious similarities to those neo-Aristotelian understandings of hylomorphic compounds that do not take the substantial form (or structure) of a hylomorphic compound to be simply another mereological part, but rather see it as the "unifying principle of the whole" that "re-identifie[s]" the parts within the substance (Marmodoro 2013, 15). Crudely put, we might think of the joint meaning as the substantial form and the subsidiary particulars as the appropriate matter in which the form is instantiated.

Polanyi's descriptions of the structure of tacit knowing often allow for the substitution of "joint meaning" for "comprehensive entity" and vice versa (*SOM* 29, *TD* 13, FR 239–240). The terms can be used, in essence, as synonyms.[17] I think that this is instructive in at least two ways. First, since the term *comprehensive entity* is often treated by Polanyi in the same way a complete hylomorphic compound (or Aristotelian substance) is by neo-Aristotelians, we get a sense of its real substantiveness as it subsists in the subsidiary particulars of which it is composed. At the same time, however, the connotations of *joint meaning* direct our attention towards the Aristotelian understanding of substantial form, the unifying principle or structure of a hylomorphic compound, which, for most hylomorphists, is not substantive on its own and cannot be used to properly refer to the complete entity. Thus while neo-Aristotelians emphasize the incompleteness of matter without form, they do not use "substantial form" and "hylomorphic compound" as synonyms.

Polanyi's usage, I would argue, has a certain Platonic resonance, since it equates the entity as a whole (including all its parts) more fully with the principle of meaningful integration of the subsidiary parts. Thus Polanyi

[17] See, for instance, "Faith and Reason" 239–240, where Polanyi uses the words "thing" and "whole" interchangeably in reference to the object of our knowledge conceived of as the *gestalt* comprised of subsidiary details before introducing the term "comprehensive entity." In *Meaning* (34–35) parallel passages use the terms "focal target" and "object of focal attention" before stating that

> The subsidiaries of from-to knowing bear on a focal target, and whatever a thing bears on may be called its meaning. Thus the focal target on which they bear is the meaning of the subsidiaries. (Ibid.)

He goes on to equate the "complex entity" made up of subsidiaries with the "joint meaning" of those subsidiaries (*M* 53–54). Elsewhere (e.g. *TD* 13 or *SOM* 29), he uses "joint meaning" in nearly identical formulations.

emphasizes, rhetorically at least, the metaphysical independence of form more forcefully than do neo-Aristotelians.

Despite this difference in emphasis, neo-Aristotelian consideration of substantial form is a rich resource that can help us to better understand what Polanyi was working towards with his theory of comprehensive entities, and can therefore serve as an aid in developing a more rigorous metaphysical vision based on Polanyi's post-critical project. To that end I shall now explore some contemporary hylomorphist insights congenial to making Polanyi's metaphysics more perspicuous. In what follows we can read Aristotelian terms such as "principle of unity," substantial form," or "structure" as analogous to the Polanyian system of *joint meaning* with the understanding that, for Polanyi, one can justifiably substitute *comprehensive entity* for *joint meaning*.

Some hylomorphists such as Kit Fine (1999) and Kathrin Koslicki (2008) advocate a mereological hylomorphism in which the substantial form is considered to be an additional part of the substance, an approach clearly out of accord with Polanyi's understanding of comprehensive entities (after all, the joint meaning is *not* simply another subsidiary detail). Meanwhile other hylomorphist theorists such as Marmodoro, Koons, and Jaworsky propose non-mereological approaches in which "form is *not a further element* in the whole on a par with the elements that make up a substance" (Marmodoro 2013, 7, emphasis original). It is, as mentioned above, a principle of unity that re-identifies parts within the context of the whole. Furthermore, as Mark Johnston points out, such principles of unity can be more or less static or more or less dynamic (2006, 664). This understanding harmonizes more seamlessly with Polanyi's theoretical contributions, specifically with his concept of joint meaning.

Like neo-Aristotelians, Polanyi emphasizes the crucial importance of the recognition of structure for the operations of science. In fact, he explicitly advocates an Aristotelian approach to science (*PK* 357–358), for instance in his insistence on the necessity of teleological structures for the study of biology, arguing forcefully that a Laplacean demon would be unable to distinguish entities such as frogs from within the undifferentiated mass of data points given by the complete map of physical particles since the identification of a frog *qua* frog depends upon the non-particulate structure of *frogginess* (Ibid., 342). Likewise, Jaworski observes that

> …scientific appeals to structure make a serious ontological demand. The most direct way of meeting that demand is to take scientific appeals to structure at face value—to claim that structure really exists, and that it plays the theoretical roles those appeals imply. We can express those roles with some slogans:
>
> *Structure matters:* it operates as an irreducible ontological principle, one that accounts at least in part for what things essentially are.
>
> *Structure makes a difference:* it operates as an irreducible explanatory principle, one that accounts at least in part for what things can do, the powers they have.
>
> *Structure counts:* it explains the unity of composite things, including the persistence of one and the same living individual through the dynamic influx and efflux of matter and energy that characterize many of its interactions with the wider world. (Jaworski 2016, 3 emphasis original)

I believe that these slogans unproblematically capture important elements of what Polanyi is moving toward when he speaks of joint meanings – they matter, they make a difference, and they count. In other words, what Polanyi means by *joint meaning* is closely related to *structure.*

Jawòrsky's approach to hylomorphism, however, often seems to gesture toward some sort of closure under physics, as long as physics properly takes structure into account. In contrast, this is not the case for Polanyi, who, as we shall see in the following chapter, affirms the existence of comprehensive entities that have no physical parts whatever.[18] Thus for Polanyi, physics – even an appropriately metaphysically reconstituted physics – will not be the source for the best descriptions of reality *per se*, but only for *physical* reality. Polanyi's position here, however, is not necessarily out of accord with the more general tenor of neo-Aristotelianism.

---

[18] I recognize that this claim may seem premature and unsupported to some interpreters of Polanyi. I shall argue my case soon enough.

For instance, Simpson suggests that "after quantum mechanics...philosophers need no longer treat the causal closure of the microphysical domain as sacrosanct" (2023, 39). Marmodoro notes that for Aristotle, the unity of a substance "derives from the metaphysical unity attained in the hylomorphic compound" (2013, 7), and that the "form is one, but a composite, and it is responsible for the oneness of the material parts of a substance by en-forming them" (Ibid., 8). She argues that this "fusion of matter and form is metaphysical and...must be explained" (Ibid., 10).

For Polanyi too, the fusion of the subsidiary particulars in and by the joint meaning is metaphysical, but its fusion won't be able to be exhaustively explained due to the tacit dimension of our epistemological situation. The metaphysical explanation we give will thus always be partial and analogical, neither complete nor literal, but such theorizing is well worth our time for all that. This post-critical understanding of the nature of metaphysical explanation strikes a tone quite different from the analytic, Anglo-American tenor of much contemporary hylomorphist theorizing, perhaps unsurprising given the genealogical influence on that sphere's evolution by the analytic tradition's past (and often present) positivist, objectivist, and scientistic pretensions.

This post-critical gloss aside, Marmodoro's reading of Aristotelian substantial form above aptly describes the functioning of joint meanings in Polanyi's metaphysics. Marmodoro, however, takes her hylomorphism in two directions problematic for post-critical thinkers: First, she argues that once "en-formed" and "re-identified," the parts of a hylomorphic compound are no longer identifiable as such, as they "have *no distinctiveness* in the substance" given that "if they were severed from the whole, they would lose [the] functional identity...conferred to them...on the basis of their role in the whole substance" (Ibid., 15). Such disconnected components would thus lose their form and become, on Marmodoro's account, something like prime matter.

From a Polanyian perspective the No Distinctiveness Doctrine outlined above is a problematic overstatement given the metaphysical closeness of knowing and being. The epistemological side of this coin holds that the parts are indistinguishable in their integration while focal attention is maintained on the comprehensive entity or joint meaning. However, since we can oscillate between focal and subsidiary awareness, shifting our focus to the part as an entity unto itself, this lack of mereological distinctiveness is not a permanent or even semi-permanent property of the part, but is rather a question of shifting relationality. This, then, is the metaphysical side of the coin – comprehensive entities can be incorporated as parts into other and new comprehensive entities without losing their substantive ontological status. When looked at through this lens, Polanyi's use of the term *integration* seems superior to Marmodoro's use of *fusion*, since the latter rings dubious notes of permanent dissolution into the larger substance.

It must be clearly noted that Marmodoro's No Distinctiveness Doctrine, also supported by Oderberg (2005) and identified as *Thomistic hylomorphism* by Johnston (2006), is not a necessary component of contemporary hylomorphism and is rejected by a number of neo-Aristotelian thinkers such as Koons, Johnston, Simpson, and Jaworsky.

There is, however, a second problematic implication of Marmodoro's hylomorphism[19] that is not so easily escaped within the neo-Aristotelian framework. To wit, the insistence that uninstantiated universals are not substantive entities. For example, she argues that

> although the way things acquire new 'form' is through physical processes which do not involve any such entity as a form (e.g. in that way a statue requires its form by the sculptor) the acquisition of a 'form' has an explanatory value which is cashed out in the metaphysical domain. (Marmodoro 2013, 10)

On Marmodoro's account, which I believe holds in general for neo-Aristotelian thought, since the substantial form is not just another mereological part of a hylomorphic compound but is rather its unifying principle, it is not an empirical entity that can be found just laying around in the sensible world. That is to say that the substantial form is not itself a complete hylomorphic compound and so is not a primary substance in the Aristotelian sense as defined in the *Categories*. Nonetheless, there it is, working its magic – mattering, making a difference, and counting, to use Jaworsky's formulation. Yet to matter, to make a difference, to count is to

[19] Marmodoro has also written perceptively on certain aspects of Plato's metaphysics (2021).

exercise power in the Eleatic sense and thus to produce IFMs, the key mark of real entities for Polanyi. Marmodoro's description seems to downplay the Eleatic power of form to shape reality, even in potency, a topic we will return to when we examine Polanyi's field theory in some detail in Chap. 4.

Polanyi recognizes a metaphysical hierarchy to be sure, but it is a hierarchy of greater or lesser richness in terms of the Eleatic Principle and IFM production (which are relational); not a yes/no hierarchy of non-substances and substances as in the contemporary analytic Aristotelian tradition. Neo-Aristotelians are of course committed to the reality of form and structure – arguably it is the very core of their positive project, however this fundamental, irreducible element of reality inevitably undergoes ontological demotion at the end of the day in accord with Aristotelian prohibitions on uninstantiated universals as fully substantive entities.

Polanyi's view, as I will argue in detail in the next two chapters, more naturally aligns with forms of Platonism that do not dismiss the importance of the sensible and material since (A) he affirms the full, substantive ontological status of immaterial or abstract entities; (B) he defines universals as the joint meaning of a class of participating particulars (i.e. comprehensive entities); and (C) he gives the potential full ontological status as real given the Eleatic shaping power it exercises through the operation of the fields he theorizes.

Even so, while the disagreement over universals is an important one, it should not overshadow the commonalities between Polanyi and neo-Aristotelianism. Certain formulations of an ontology of powers worked out within the context of Robert Koons' "staunch hylomorphism," for instance, are quite congenial to much of Polanyi's project:

> Powers are individuated by the character of their exercise—that is, by the character of the outcomes they naturally produce. Difference in teleological properties corresponds to such a difference in outcomes...This picture clearly differentiates hylomorphism from any version of materialism, since it implies that every material element has a fundamental nature that anticipates...all of the possible kinds of composite substance in which an element of that kind could be incorporated. (2014, 168)

Here we see echoes of Polanyi's work on dual control and boundary conditions, a resonance also seen in Jaworsky's work when he notes that

> The hylomorphic view is also committed to property pluralism. It implies that structured individuals have properties of at least two sorts: properties due to their individual-making structures (or their integration into individuals with individual-making structures), and properties due to their materials alone, independent of the ways they are structured. (2016, 106)

Like Polanyi, neo-Aristotelians are engaged in a struggle to replace and reform the cramped and reductionistic metaphysical perspective that has permeated and tended to define the modern, critical project. Neo-Aristotelianism arguably faces some significant rhetorical headwinds given the fact that most of its theoreticians ply their trade within a tradition of analytic discourse that has arguably been constituted by the very metaphysical view it seeks to shake off. Central to the neo-Aristotelian project in this regard is a wave of constructive efforts to rehabilitate the metaphysical doctrine of hylomorphism as a live option which can not only be constructively merged with contemporary science, but may in fact be able to foster further creative scientific endeavor. To the extent the neo-Aristotelian project is successful in these central metaphysical aims, it can serve as an important partner in the post-critical project as identified by Polanyi.

In the brief and preliminary explorations I've offered above, I've attempted to begin to flesh out some constructive points of contact between Polanyi's legacy and the field of neo-Aristotelian realism. The literature produced by these contemporary hylomorphist thinkers offers a rich set of resources for future refinement of the post-critical metaphysics I will be exploring and arguing for in the following chapters, since, as Johnston notes, "hylomorphism provides some philosophical apparatus that enables us to get to the base of some of our unreflective thoughts about complex wholes" (2006, 663). This is so because such a post-critical metaphysics will be based to a significant extent on Polanyi's understanding of complex wholes, which is to say, comprehensive entities.

4 Chapter Conclusion

Polanyi's epistemological project is deeply intertwined with his particular brand of metaphysical realism – a realism which affirms our ability to make contact with and grasp reality, if only partially, by means of both philosophical reflection on first principles and empirical observation. Polanyi self-consciously modelled his realist position on that of Copernicus and his followers and I have thus dubbed it a Copernican realism.

Interpreters of Polanyi have differed in certain respects over precisely how to characterize his realism, however. I have attempted to trace these debates and have given especial attention to the ontological distinction urged by Walt Gulick (following in some respects Harry Prosch) between the meaningful and the real, a distinction I reject as re-introducing the fact-value distinction that Polanyi worked to abolish.

Following on this I have attempted to bring Polanyi's Copernican realism into conversation with three contemporary metaphysical realisms. These forays have demonstrated, I believe, that there is significant overlap between Polanyi's post-critical metaphysics and contemporary discussions on both the analytic and continental sides of the philosophical divide.

While Polanyi would be highly critical of ontic structural realism's constrictive and stipulative Principle of Natural Closure, there would still be room for constructive dialogue with regard to its embrace of the scale-relative ontology that facilitates its inclusive rainforest realism.

Convergence is greater, however, when we come to Graham Harman's object-oriented ontology, whose emphasis on a flat ontology of objects will lend itself quite well in many respects to a more robust development of Polanyi's theory of comprehensive entities.

Likewise, a post-critical metaphysics based on Polanyi's theorizing will be able to make much common cause with neo-Aristotelianism and its attendant contemporary hylomorphism, especially with regard to resonances between joint meanings (which define and determine the structure of comprehensive entities on Polanyi's account) and substantial forms (which order and structure suitable matter into Aristotelian substances on a hylomorphic account).

As we proceed to immerse ourselves more fully in a Polanyian metaphysics, we will see how some of these resonances play out in more detail.

References

Aristotle. *De Anima* trans. J.A. Smith. *Internet Classics Archive.* Accessed September 28, 2021. http://classics.mit.edu/Aristotle/soul.html.

_____. *Categories* trans. E.M. Edghill. *The Internet Classics Archive.* Accessed March 10, 2023. http://classics.mit.edu/Aristotle/categories.1.1.html.

Cannon, Dale. 1999–2000. "Some Aspects of Polanyi's Version of Realism." *Tradition and Discovery* 26/3: 51–61.

Dennett, Daniel C. 1991. "Real Patterns." *The Journal of Philosophy* 88/1: 27–51.

Fine, Kit. 1999. "Things and Their Parts." *Midwest Studies in Philosophy.* Vol. 23: 61–74.

Fujisawa, Norio. 1974. "Εχειν, Μετέχειν, and Idioms of 'Paradeigmatism' in Plato's Theory of Forms." *Phronesis* 19/1: 30–58.

Gerson, Lloyd P. 2005. Aristotle and Other Platonists. Ithaca, NY: Cornell University Press.

Gulick, Walter B. 1999–2000. "Beyond Epistemology to Realms of Meaning." *Tradition and Discovery* 26/3: 24–41.

———. 1999. "The Meaningful *and* the Real in Polanyian Perspective." *Polanyiana.* 8/1–2: http://www.kfki.hu/chemonet/polanyi/; http://www.ch.bme.hu/chemonet/polanyi/

Harman, Graham. 2002. *Tool-Being: Heidegger and the Metaphysics of Objects.* Chicago, Il: Open Court Books.

———. 2007. "On Vicarious Causation." *Collapse II*: 187–221.

———. 2010. "I am also of the opinion that materialism must be destroyed." *Environment and Planning D: Society and Space*, Volume 28: 772 – 790.

———. 2012. *Weird Realism: Lovecraft and Philosophy.* Washington: Zero Books.

———. 2018. *Object-Oriented Ontology: A New Theory of Everything.* Pelican Books.

Jaworski, William. 2016. *Structure and Metaphysics of Mind: How Hylomorphism Solves the Mind-Body Problem.* Oxford: Oxford University Press.

Jha, S.R. 1999–2000. "Polanyi's Problematic 'Man in Thought': the Tacit and the Real – an Exploration and a Critique." *Tradition and Discovery* 26/3: 15–23.

Johnston, Mark. 2006. "Hylomorphism." *The Journal of Philosophy* 103/12. Special Issue: Parts and Wholes: 652–698.

Koons, Robert C. 2014. "Staunch vs. Faint-Hearted Hylomorphism: Toward an Account of Composition." *Res Philosophica* 91/2: 151–177.

Koslicki, Kathryn. 2008. *The Structure of Objects.* Oxford: Oxford University Press.

Ladyman, James. 1998. "What is Structural Realism?" Studies in History and Philosophy of Science Part A. 29/3: 409-424.
Ladyman, James, Ross, Don, and Collier, John G. 2007. *Every Thing Must Go: Metaphysics Naturalized.* Oxford: Oxford University Press.
Lowney II, Charles. 2008. "The Tacit in Frege." Polanyiana. 1-2/17: 19–37.
Marmodoro, Anna. 2013. "Aristotle's Hylomorphism without Reconditioning." *Philosophical Inquiry* 37/1–2: 5–22.
Meek, Esther Lightcap. 2017. *Contact with Reality: Michael Polanyi's Realism and Why It Matters.* Eugene, OR: Cascade Books.
Meillasoux, Quentin. 2006. *After Finitude: An Essay on the Necessity of Contingency.* Trans. Ray Brassier. New York: Bloomsbury.
Mullins, Phil. 1999–2000. "The Real *As* Meaningful." *Tradition and Discovery* 26/3: 42–50.
———. 1997. "Polanyi's Participative Realism." *Polanyiana* 6/2: https://www.kfki.hu/~cheminfo/polanyi/9702/mullins.html
Oderberg, David S. 2005. "Hylemorphic Dualism." *Social Philosophy and Policy* 22/2: 70–99.
Plato. 1997. *Sophist* trans. Nicholas P. White, in *Plato: Complete Works.* Edited by John M. Cooper. Indianapolis, IN: Hackett Publishing Company.
Polanyi, Michael. 1961. "Faith and Reason." *Journal of Religion* 41/4: 237–247.
———. 1962 [1958]. *Personal Knowledge: Towards a Post-Critical Philosophy.* Chicago, IL: University of Chicago Press.
———. 1964 [1946]. *Science, Faith, and Society.* London: Oxford University Press.
———. 1964. "Lecture 1: The Metaphysical Reach of Science." *Duke Lectures.* Delivered at Duke University, February 10, 1964. http://www.polanyisociety.org/Duke/Duke1.pdf
———. 1965a. "Lecture 1: Science and Reality." *Wesleyan Lectures.* Delivered at Wesleyan University, September 29, 1965. http://www.polanyisociety.org/WesleyanLectures/Weslyn-lec1%2D%2D9-29-65-searchable.pdf
———. 1965b. "Lecture 3: The Creative Imagination." *Wesleyan Lectures.* Delivered at Wesleyan University, October 31, 1965.http://www.polanyisociety.org/WesleyanLectures/Weslyn-lec3-10-21-65.pdf
———. 1966. *The Tacit Dimension.* Chicago, IL: University of Chicago Press.
———. 1967. "Science and Reality" *The British Journal for the Philosophy of Science* Vol. 18, No. 3: 177–196.
———. 2013 [1959]. *The Study of Man.* New York, NY: Routledge.

Polanyi, Michael and Prosch, Harry. 1975. *Meaning*. Chicago, IL: University of Chicago Press.
Prosch, Harry. 1972 "Polanyi's Ethics." *Ethics.* 82/2: 91–113.
Puddefoot, John C. 1993–1994 "Resonance Realism." *Tradition and Discovery* 20/3: 29–39.
Rea, Michael C. 2011. "Hylomorphism Reconditioned." *Philosophical Perspectives* 25/1: 341–358.
Sanders, Andy F. 1999–2000. "Science, Religion, and Polanyi's Comprehensive Realism." *Tradition and Discovery* 26/3: 84–95.
Simpson, William M.R. 2023. *Hylomorphism.* Cambridge: Cambridge University Press.
Simpson, William M.R., Koons, Robert, and Teh, Nicholas J. 2017. *Neo-Aristotelian Perspectives on Contemporary Science.* New York: Routledge.
Veal, Damian. 2012. "Who's Afraid of Scientism?: Interview with James Ladyman." *Collapse V*: 135–185.
Worrall, J., 1989. "Structural Realism: The Best of Both Worlds?" *Dialectica* 43: 99–124.

From Epistemology to Metaphysics

1 Polanyi's Unfinished Metaphysics

As we have seen, Polanyi's entire post-critical project is bound up tightly with his commitment to what I have called Copernican realism, a metaphysical thread which runs from his early philosophical writings (such as 1946's *Science, Faith, and Society*) through his later, more mature philosophical output (1967's "Science and Reality"). This metaphysical commitment to a realism that affirms both philosophical reasoning about first principles *and* empirical investigation as valid methods for investigating reality provides a crucial context within which he works out a fairly systematic epistemology. His metaphysics, meanwhile, does not receive the same degree of systematic attention, though, as Phil Mullins (1997, 1999–2000, 2006–2007, 2018, and 2022) points out, Polanyi does revisit metaphysical questions periodically – especially in the Duke and Wesleyan lectures which culminate in "Science and Reality," as we have already seen, but also in *The Tacit Dimension* as well as in certain passages of the earlier work, *Personal Knowledge*. While the later development undoubtedly gives us a more robust set of metaphysical doctrines than

M. E. Turkis II, *The Metaphysics of Michael Polanyi*,
https://doi.org/10.1007/978-3-031-41211-0_4

are evident in the earlier texts (including his work on emergence),[1] Polanyi never comes close to giving us a comprehensive metaphysical theory that would undergird his theory of knowing, and this is noteworthy given the closely intertwined relationship in his thought of knowing and being.[2]

In this chapter we will examine Polanyi's unfinished metaphysics before proceeding in the following chapters to argue that his metaphysical project needs to drink deeply from a certain tradition of Platonism if it is to remain coherent and defensible. To begin with, we will look at the ways in which Polanyi's work tacitly affirms the priority of metaphysics over epistemology, a priority more characteristic of ancient thinkers, before moving on to examine what Polanyi calls in *The Tacit Dimension* the ontological aspect of tacit knowing. Following that we will more fully examine comprehensive entities, a category central to Polanyi's epistemology as well as his metaphysics. We will see more clearly here that despite his emphasis on a certain type of metaphysical hierarchy (which I will continue to argue is based on the Eleatic Principle and IFM production), Polanyi's teachings on comprehensive entities imply a flat ontology of comprehensive entities inclusive of both the material and immaterial within which hierarchies of complexity can unfold. I will argue, furthermore, that this ontological outlook can profit from some of the language and insight generated by contemporary Aristotelian hylomorphism as well as Object-Oriented Ontology.

---

[1] Polanyi explicitly develops a version of emergence theory and also, as I will go on to argue, affirms the existence of ordering principles that guide emergence, a combination that is not likely to be acceptable to all emergence theorists, though it will be, no doubt, for some (for more on this see Jon Fennell's "Is Polanyi's Emergence Reductive?" (2017). It is thus unsurprising that some interpreters of Polanyi's work, such as Lowney and Gulick, for example, emphasize emergence as of central importance, while others (e.g. Mullins, Cannon, and Meek) do not. I am myself of the camp that does not find Polanyi's work on emergence theory to be among his most important contributions. In light of this, as well as of the fact that debates over emergence can be highly technical and involved, I will not engage with emergence theory in any sustained way, though it will be touched on from time to time. I imagine that this will frustrate some readers and elate others.

[2] See Gallardo (2003) pages 297 and 301 for further comments on the lack of systematicity in Polanyi's work.

2 The Priority of Metaphysics

Given that Polanyi's main philosophical contribution is his theory of knowledge, my contention that he, like the ancients, prioritizes metaphysics over epistemology may seem strange. For instance, Phil Mullins (2006–2007), following Marjorie Grene (1977), sees Polanyi as approaching metaphysics principally as an outgrowth of his epistemology.

I willingly concede that Polanyi is not always clear about what I take to be his prioritization of the metaphysical and that it is often a tacit implication of his thought rather than an explicit doctrine. Nonetheless, in *The Tacit Dimension* Polanyi notes that he ventured into philosophy in large part due to his realization that the explicit philosophy of his contemporary scientific colleagues "left science defenseless against the Soviet doctrine" (81) that made Communist Party doctrine the ultimate arbiter of scientific truth, a position which led to tragedies such as the Lysenko affair.[3] This concern led Polanyi "to decide that only on metaphysical grounds can we account for the intrinsic powers of human inventiveness" (Ibid.). Likewise, in *Meaning* he claims that "all our knowledge of the external world actually rests on tacitly accepted metaphysical grounds" which give rise to an "inexhaustible profundity, because what we perceive is tacitly understood by us to be an aspect of reality, and aspects of reality are tacitly believed to be clues to boundless undisclosed, and perhaps yet unthinkable, experiences" (188).[4] These core metaphysical beliefs are the essence of Polanyi's Copernican realism, previously discussed. They "necessarily assure discipline and foster originality in science," (Ibid., 189), reconcile the "apparent self-contradiction" posed by "the constitutive coexistence in science of authority and dissent" (*TD* 68) and are also the "metaphysical grounds of intellectual life in a free, dynamic society: the principles which safeguard intellectual life in…a society of explorers" (Ibid., 83).

[3]Trofim Lysenko was a Soviet scientist who rejected Mendelian genetics and proposed Lamarckian theories that would purportedly increase crop yields. Lysenko, with Stalin's support, was able to rhetorically position his theories – already known to be false on the basis of the settled genetic science of the time – as in line with Communist Party doctrine and thus *true*. Some of Lysenko's Mendelian critics were imprisoned or executed, and crop yields suffered (Witkowski 2008).

[4]See also *The Tacit Dimension*, pages 67–68.

At the end of *Personal Knowledge* Polanyi says that the book's purpose has been to re-orient us toward reality by

> …re-equip[ing] men with the faculties which centuries of critical thought have taught them to distrust…[and to] contemplate thus a picture of things restored to their fairly obvious nature…For once men have been made to realize the crippling mutilations imposed by an objectivist framework…many fresh minds will turn to the task of reinterpreting the world as it is, and as it then once more will be seen to be. (381)

This is clearly a metaphysical goal, although *PK* sets out to achieve it by launching a post-critical epistemology that will allow us to openly countenance these metaphysical grounds. Being begets knowing, and knowing begets being. The two are inextricably tangled and no non-circular justification is to be had for either the one or the other, though this is not troubling once one has crossed fully into post-critical territory. Thus Polanyi proclaims at the outset of *PK* that

> To say that the discovery of objective truth in science consists in the apprehension of a rationality which commands our respect and arouses our contemplative admiration; that such discovery, while using the experience of our senses as clues, transcends this experience by embracing the vision of a reality beyond the impressions of our senses, a vision which speaks for itself in guiding us to an ever deeper understanding of reality—such an account of scientific procedure would be generally shrugged aside as *outdated Platonism*: a piece of mystery-mongering unworthy of an enlightened age. Yet it is precisely on this conception of objectivity that I wish to insist… (PK 5, emphasis added)

As this project continues to unfold, I hope to offer richer and more fleshed-out account of an updated post-critical Platonism. Yet before doing so it behooves us to note that another side of the inseparability of metaphysics and epistemology is the inseparability of phenomenology and metaphysics. Consider the following passage from *The Study of Man*, which reiterates once more the metaphysical end-game of *PK*'s epistemological method:

> Modern science disclaims any intention of understanding the hidden nature of things; its philosophy condemns any such endeavour as vague, misleading and altogether unscientific. But I refuse to heed this warning. I agree that the process of understanding leads beyond—indeed far beyond—what a strict empiricism regards as the domain of legitimate knowledge; but I reject such an empiricism. If consistently applied, it would discredit any knowledge whatever and it can be upheld only by allowing it to remain inconsistent. It is permitted this inconsistency because its ruthless mutilation of human experience lends it such a high reputation for scientific severity, that its prestige overrides the defectiveness of its own foundations. Our acknowledgment of understanding as a valid form of knowing will go a long way towards liberating our minds from this violent and inefficient despotism. (20)

We can discern here that the priority of metaphysics is urged on largely phenomenological grounds.[5] For Polanyi approves heartily of the groping about beyond the screen of appearances as entailed by the integrative force of *understanding* or *comprehension* (over and against a third-person, objectivist and propositional conception of knowledge hemmed in by positivist or semi-positivist pieties). Yet the grounds for his affirmation of such understanding is ultimately a phenomenological move since it is based on partial yet compelling descriptions of the experience of comprehension that we all resonate with viscerally, but which of course would be bowled over as so much first-person balderdash by a critical, third-person conception of knowing. By doubling down on our ordinary and commonsensically indubitable experiences of understanding as more fundamental than and superior to the critical canons established by the epistemological turn in modern thought, Polanyi throws in his lot with the Attic conception of the fitted-ness of mind to world and vice versa. In making this move the line between phenomenology and metaphysics is blurred.

To give a more concrete example of the entanglement of metaphysics, epistemology, and phenomenology, let us consider the way science is pursued. Polanyi argues, as we have seen, that "the scientist can conceive problems and pursue their investigation only by believing in a hidden

[5] Though in contrast to Husserl, Polanyi does not bracket the transcendent, but embraces it.

reality on which science bears" (*TD* 74–75). This, according to Polanyi, is the metaphysics of all practical science. But the actual practice of the scientist is experiential through and through, and thus phenomenological, since only on the basis of various types of experience are scientific results obtained. Such scientific findings make up the body of knowledge that is the evolving scientific consensus at any given time, a body of knowledge which is always reaching beyond the bounds of collective scientific experience by means of the faith placed in results obtained in overlapping areas of expertise as well as by means of new hypotheses which reach beyond the current state of experience and which, in physics, sometimes reach beyond all possible experience into the realm of metaphysical speculation (the postulation of the multiverse is one such example). This scientific consensus, simultaneously flowing out of phenomenological experience and reaching beyond it, is governed by "standards of plausibility…standards of beauty and originality," (Ibid., 72) standards which only have authority by virtue of the tacit metaphysical commitment of those who submit to them. Thus, "the tradition of science induces its own renewal by bearing on a reality beyond experience," (Ibid., 74) yet it can only do this by giving metaphysical weight to the phenomenological experience of scientists.

By this road Polanyi arrives at the concept of *universal intent*, which is an acknowledgement of the inescapable metaphysical weight of experience for which no non-circular justification is available, since "the scientist, having relied throughout his inquiry on the presence of something real hidden out there, will justly rely on that external presence also for claiming the validity of the result that satisfies his quest" and will thus "claim that his results are universally valid" (Ibid., 78).

Thus even though "modern science arose claiming to be grounded in experience and not on a metaphysics derived from first principles," (Ibid., 70) the growth of science and the experience and expectations of scientists can themselves only be taken seriously on metaphysical grounds residing somewhere in the vicinity of Polanyi's Copernican realism. Such a metaphysics is thus not a purely abstract or other-worldly affair, but is really quite ordinary and practically-rooted, giving rise to concrete social arrangements:

> All institutions serving the advancement and dissemination of science rely on the supposition that a field of potential systematic progress exists, ready to be revealed by the independent initiative of individual scientists. It is in view of this belief that scientists are appointed for life to the pursuit of research and permanent subsidies are granted to them for this purpose. Many expensive buildings, pieces of equipment, journals, etc. are founded and maintained in this belief. It is the most general traditional belief which a novice joining the scientific community accepts in becoming a scientist. (Ibid., 71)

The institutions of science are not unique in this regard. A great many, and perhaps in some sense all, human endeavors that become formally or informally institutionalized share some version of the idea of a field of potential systematic progress.

Yet due consideration of the ordinary, everyday nature of metaphysics will lead us naturally, argues Polanyi, to affirm metaphysical entities that are immaterial and abstract:

> …our heuristic cravings imply, like our bodily appetites, the existence of something which has the properties required to satisfy us, and that the intimations which guide our striving express this belief. But the satisfier of our craving has in this case no bodily existence; it is not a hidden object, *but an idea never yet conceived.* We hope that as we work at the problem this idea will come to us, whether all at once or bit by bit; and only if we believe that this solution exists can we passionately search for it and evoke from ourselves heuristic steps towards its discovery. Therefore, as it emerges in response to our search for something we believe to be there, discovery, or supposed discovery, will always come to us with the conviction of its being true. It arrives accredited in advance by the heuristic craving which evoked it. (*PK* 129, emphasis added)

We see here again the way in which Polanyi's heuristic phenomenology is ultimately a metaphysics of Eros, the force that propels us toward discovery. And, like Platonism, it affirms the existence of uninstantiated ideas which act as attractors, though this is to run ahead of our argument a bit.

We entertain and follow up on heuristic cravings within the context of our lives as personal centers of achievement shaped by our calling, which is "determined by our innate faculties and our early upbringing within our own culture" (*PK* 379).

3 The Ontological Aspect of Tacit Knowing

I have been arguing that, despite his renown as an epistemologist, Polanyi's philosophical program is fundamentally metaphysical. Aside from the writings on metaphysical realism examined earlier, Polanyi is perhaps clearest about this in *The Tacit Dimension* where he outlines four aspects of the structure of tacit knowing: The *functional aspect,* the *phenomenal aspect, the semantic aspect*, and *ontological aspect.*

The *functional aspect* explains "how we come to know [subsidiary] particulars, without becoming able to identify them…*we know the first term only by relying on our awareness of it for attending to the second"* (TD 10, emphasis original). The *phenomenal aspect* outlines the phenomenological structure of our from/to attentional structure, noting "that we are aware of the proximal term of an act of tacit knowing in the appearance of its distal term; we are aware of that *from* which we are attending *to* another thing, in the *appearance* of that thing" (Ibid., 11, emphasis original).

Polanyi notes that *meaning* is key to understanding the third and fourth aspects of tacit knowing. He defines it as "significance in the relation of the two terms of tacit knowing which combines its functional and phenomenal aspects" (Ibid.). By way of example, he notes that "when the sight of certain syllables makes us expect an electric shock, we may say that they *signify* the approach of a shock. This is their *meaning* to us" (Ibid.). The *semantic aspect* of tacit knowing denotes the way that "an interpretative effort transposes meaningless feelings into meaningful ones, and places these at some distance from the original feeling," as when "we become aware of the feelings in our hand in terms of their meaning located at the tip of the probe or stick to which we are attending" (Ibid., 12–13). Thus, "all meaning tends to be displaced *away from ourselves*" (Ibid., emphasis original).

Before too long I will further develop Polanyi's conception of meaning in ways that will show how it is integral also to understanding the *ontological aspect* of tacit knowing, which is our primary interest here. In affirming the ontological aspect Polanyi openly and explicitly affirms the structural metaphysical connection between knowing and being. In *Personal Knowledge* he prefigures the ontological aspect when he notes that "every comprehension of a whole acknowledges the reality of it; and whatever we comprehend both means something to us and to a certain extent at least means something also existentially, in itself" (344). In *The Tacit Dimension* he argues that

> From the three aspects of tacit knowing that I have defined so far—the functional, the phenomenal, and the semantic – we can deduce a fourth aspect, which tells us what tacit knowing is a knowledge of. This will represent its *ontological* aspect. Since tacit knowing establishes a meaningful relation between two terms, we may identify it with the *understanding* of the comprehensive entity which these two terms jointly constitute. Thus the proximal term represents the *particulars* of this entity, and we can say, accordingly, that we comprehend the entity by relying on our awareness of its particulars for attending to their joint meaning. (*TD* 13)

The tacit integration of subsidiaries into a totality grasped by an act of knowing thus renders epistemological access to an ontological structure – that of the comprehensive entity, a topic we will explore in much depth presently. Yet this structure, according to Polanyi, is not merely a mental construction nor a phenomenal mirage. He goes on to say that

...what it is that we know by understanding a comprehensive entity, makes an *ontological* reference to it...It seems plausible then to *assume in all other instances of tacit knowing the correspondence between the structure of comprehension and the structure of the comprehensive entity which is its object.* And we would expect then to find the structure of tacit knowing duplicated in the principles which account for the stability and effectiveness of all real comprehensive entities. (*TD* 33–34, emphasis original).

This, then, is where Polanyi's Copernican realism, affirming of both philosophical and empirical approaches to reality openly affirms the metaphysical fitted-ness of mind to world. If this move be thought by some naïve, Polanyi responds by arguing that the progress and validity of science itself (as well as all other disciplines) is wholly dependent upon just this sort of metaphysical assumption, whether this fact is openly recognized or not, since any participant in the process of scientific discovery must affirm that she holds some scientific knowledge which can serve as a basis for further discovery, and to do so is to reveal oneself to be "deeply committed to the conviction that there is something there to be discovered," and to affirm in oneself "a compelling sense of responsibility for the pursuit of a hidden truth" (Ibid., 25). This move is transparently metaphysical as well as empirical and methodological since it demands that we make a "personal judgment in relating evidence to an external reality, an aspect of which [we are] seeking to apprehend" (Ibid.).

The discoverer (and we are all discoverers on Polanyi's view) is working to comprehend *an aspect* of reality, not the whole of it. And Polanyi explicitly recognizes that "the anticipation of discovery, like discovery itself, may turn out to be a delusion" (Ibid.).

The Polanyian view is therefore not one of naïve one-to-one correspondence but rather involves a forthright acceptance of the fallibility of our ability to track reality that does not fall into nihilist pessimism. Polanyi recognizes openly that the metaphysical connection between knowing and being is indemonstrable in purely objective, non-circular terms, although tacit affirmation of such a connection is necessary for us to get on with things. This is, again, the heart of the post-critical project: "To find a stable alternative to [the] ideal of objectivity" (Ibid.). Knowledge is partial and fallible, but it does provide a route to reality by allowing us some degree of purchase on being.

Furthermore, the structure of knowing provides a guide to certain aspects of the structure of being. We will recall that knowledge for Polanyi has a from/to attentional structure whereby subsidiary particulars form integrative *gestalt* wholes, or comprehensive entities. The implication of Polanyi's view of the structural kinship of knowing and being, then, is that all of reality is metaphysically constituted by an ontology of comprehensive entities, an ontology which is flat in the sense that the structure of all beings is that of the comprehensive entity.

Polanyi, however, does not make these last two claims explicitly. As I have mentioned a number of times, his epistemology is quite systematic, while his metaphysics go relatively undeveloped. Polanyi works out his Copernican realism to a fair extent (though not clearly enough so as to avoid significant debate over precisely what his realism amounts to, as we have seen). He also, of course, gives the explanation of the ontological aspect of tacit knowing we have just examined. In my view this is where he is clearest and most systematic metaphysically, but it is patchy, to say the least. In what follows I will attempt to draw out some of the implications of Polanyi's commitment to the connection between knowing and being, his piecemeal metaphysics, and his epistemological teachings in order to put them in more systematic order.

4 Comprehensive Entities

Phil Mullins, who has probably written more about Polanyi's concept of the comprehensive entity than anyone else, argues that

> Polanyi came to view this locution as a very useful one that served to set forth clearly some implications of his theory of tacit knowing. In the final analysis…Polanyi's comments about the "comprehensive entity" and the ontological aspect of tacit knowing need to be viewed strictly as ways to re-articulate his claims about the active shaping of knowledge. That is, Polanyi's claims regarding the "comprehensive entity" must always be seen within the context of his claims about "comprehension." Nevertheless, what Polanyi says about "comprehensive entities" helps to sketch out Polanyi's metaphysics, which as Grene points out is a metaphysics approached through epistemology. Such a metaphysics is pluralistic; it acknowledges a rich variety of realities. (Mullins 2006–2007, 27)

Over the course of several articles (1997, 1999–2000, 2006–2007 and 2018) Mullins fleshes out a history of Polanyi's use of the term. He finds one reference to comprehensive entities in the Gifford Lectures (in the lecture entitled "Living Beings" given in 1952) (2018, 1–2). The term is used in *Personal Knowledge* (1958) in a relatively restricted sense to refer to living things and machines, but Mullins demonstrates persuasively that, beginning with *The Study of Man* (1959), Polanyi expands the range of the term's application, thinking of it as "as simply a term useful to identify any object of knowledge, living or otherwise… a focal whole…that includes (or comprehends) subsidiaries" (Ibid., 30).

According to Mullins, Marjorie Grene, herself an important philosophical partner and interpreter of Polanyi, never picked up on Polanyi's more wide-ranging use of the term, which can be found throughout *The Study of Man,* the Duke and Terry Lectures, *The Tacit Dimension*, and *Meaning*. Nonetheless Grene's tendency to continue to restrict the use of "comprehensive entity" to refer only to living things (and perhaps machines) is echoed from time to time by other interpreters of Polanyi in discussions of his work, perhaps because most readers of Polanyi do not focus in a sustained, robust way on the metaphysical implications of the term.

Despite this scattered hesitance I agree with Charles Lowney that the broader use of the term is more helpful metaphysically, especially when discussing the reality of universals in Polanyi.[6] Happily, Mullins's historical work shows that this broader understanding was the one held by Polanyi in his more mature years as a philosopher, as further citations throughout this section will reiterate.

As Mullins points out, the *comprehensive* nature of entities is tied for Polanyi to *comprehension*, which is an "unformalizable process striving towards an unspecifiable achievement…attributed to the agency of a centre seeking satisfaction in the light of its own standards" (*PK* 398).

Mullins sees this as implying that only those entities that are encountered and known (i.e. comprehended) ought to be characterized as comprehensive entities, claiming that Polanyi "makes it quite clear that it is not acceptable to think of comprehensive entities as merely things external to the knower," (Mullins 1997, IV) yet I'm not so certain. Where does Polanyi make this prohibition clear? Much depends on the meaning of

[6] Charles Lowney, email message to the author, March 7, 2021. See also Lowney (2017), where he states: "One aspect…that I have drawn attention to is how we see new possibilities for dwelling in and understanding things via universals, concepts, or 'forms.' A new conception of form or *eidos* is linked to the *through* in a *from-through-to* structure. We begin to see how various particulars can be gathered up into a coherent ideal or a comprehensive entity that is a discovery and not merely a construct" (162). For Lowney, while universals are real and exist, they may present a different type of reality since they are not coextensive with natural entities that have parts. I, however, do not introduce such a distinction.

"mere, external things." Polanyi's emphasis is clearly on *our* epistemological contact with a non-reductionist reality, to be sure, and some real things within his scheme are arguably products of sentient activity (poems, conversations, etc.), as will be laid out clearly in due course, but nowhere does he forbid thinking of comprehensive entities *per se* as having an independent existence within the world that we commonly inhabit. That said, he certainly never explicitly takes up the question of the independent existence of comprehensive entities aside from implying such independence in his explanation of the ontological aspect of tacit knowing, and this leaves us an important gap in a Polanyian metaphysics that will profit from some of the insights generated by OOO theorists like Graham Harman, as well as from contemporary hylomorphism, as we will see.

I thus see the view that takes comprehensive entities to properly apply only to known entities as unnecessarily restrictive, since we can just as well read the comprehension-to-comprehensive move (i.e. knowing-to-being) in reverse as a causal being-to-knowing relationship between the *comprehensiveness* of the entity which gives rise to its possible *comprehension* by the knower. Indeed, this reading has greater explanatory power: All entities *qua* entities, insofar as they are real, are *comprehensive* entities. As such they are *potentially knowable/comprehensible* by sentient agents, even if they are never *actually known/comprehended* by such agents. The co-extensiveness of knowing and being is thus preserved inviolate and combined with an appropriately balanced post-critical faith in and humility with regard to our epistemic powers.

The view outlined above allows me to wholeheartedly agree with Mullins when he argues that "Polanyi's term emphasizes the active and personal participation of the knower in the known" (Ibid.), even as I point out that this is often operative at the level of potentiality rather than actuality. When faced with the interconnectedness of knowing and being, the question of directionality naturally arises: Whether to accentuate the knower as bound up with and dependent upon the known for knowing or to emphasize the dependence of the known on the knowing activity? The proper post-critical response in my view is to argue that this is a two-way street, with being and knowing (i.e. *comprehension*) logically simultaneous and coincident. This is because, as will be worked out in the following section, a comprehensive entity's mereological relationship to its subsidiary particulars is defined by the entity's *comprehension* of those particulars (that is to say its indwelling of the particulars towards the fulfillment of its joint meaning). To be is to mean, then, and it is also, in a sense, to know. Thus we are left with an account that affirms the primacy of metaphysics and restores epistemology to its place as an aspect of metaphysics.

I have been arguing against taking Polanyi's epistemological focus as a tacit denial of the more foundational role of metaphysics for Polanyi's project, perhaps even going so far as to say that for Polanyi, epistemology *is* metaphysics (Lowney 2013–2014).[7] But this is in accord with the larger spirit of Mullins's work on Polanyi's participative realism which holds that "Polanyi's account is a perspective that does not make a sharp distinction between the process of understanding and the product understood" (2006–2007, 33). He rightly points to Heideggerian overtones of being-in-the-world that Polanyi himself drew attention to when characterizing his own theory of knowing/comprehension based on our indwelling of subsidiary particulars as participation in reality.

Since, on Mullins's reading, "Polanyi's…is a perspective that marks how impossible it is ontologically to separate the knower and the known," (Ibid.) he sees Polanyi as being metaphysically at home with Peirce and the medieval realists since for these thinkers, "knowing and being are woven inextricably together" with the "ontological status of entities…not tied…to 'existence' and tangibility, but to an entity's intelligibility and its prospect for greater intelligibility" (Ibid., 37). Precisely - though one might want to point out that on this view "existence" just is, to a very great degree, intelligibility (though for finite beings not total intelligibility), since on my view meaning is being. Thus Mullins holds that the

> metaphysical implications of "post-critical' philosophy fit into the metaphysical orientation of medieval realists. Polanyi, like Peirce, follows Plato

[7] In this I partially agree with Lowney (2013–2014), who claims that, *at a deep level*, epistemology is indistinguishable from ontology, but at higher, developed levels there *is a divergence* and we need a distinction between the epistemic and the ontic for "true" to be applicable. I do not necessarily think that there is a real distinction, even at the higher levels.

> in emphasizing the special human vocation as focused on inquiry or understanding. Understanding for Polanyi is "comprehension," an active shaping of knowledge by a human being who has the capacity to dwell in and tacitly integrate an extraordinary range of particulars. Humans integrate tacitly held particulars to attend to "comprehensive entities" and such entities can be viewed as skillful human performances or achievements that serve or realize self-set standards. But "comprehensive entities" are realities and some realities are more richly significant than others, according to Polanyi. Such richly real "comprehensive entities" have the capacity to be understood more deeply as human inquiry proceeds…Human beings are members of communities of inquiry and the on-going project of persons in such communities is to explore the intricacies of that encompassing, inexhaustible reality called the cosmos. (Ibid., 38)

Mullins's historical and interpretive work on comprehensive entities[8] is invaluable. He has charted the evolution of the term within Polanyi's work from a reference to living things and machines to a broader conception applicable to any entity that is known, which, when taken with Polanyi's understanding of knowing and being as co-extensive, leads him to develop his reading of Polanyi's realism as participative. Mullins's project here is one of elucidation and history. He does not attempt to theorize a post-critical metaphysics more thorough or robust than what Polanyi himself left us, and he (Mullins) notes that "Polanyi does not seem to have sorted out all of the metaphysical issues connected with his more generic use of 'comprehensive entity'" (Ibid., 3). Our exploration of Mullins's analysis of the Polanyian concept of comprehensive entities thus sets the stage for a more thorough post-critical metaphysics to be built upon the foundation of Polanyi's theory of the comprehensive entity as the go-between for knowing and being.

[8] Mullins (1999–2000, 2006–2007, 2018) also argues that Polanyi's realism unites *polyvalent* and *bodily* aspects "Comprehensive entities have a bodily dimension insofar as those who grasp such entities must dwell in subsidiaries and actively shape or coordinate their tacit integration… comprehensive entities have a polyvalent dimension insofar as such entities open up the indeterminate meaning of reality" (Mullins 2018, 5). "Bodily" here must be distinguished from "physical," since immaterial entities such as symbols can be indwelt.

5 Toward a Speculative Metaphysics of Comprehensive Entities

In girding our loins for the speculative project before us it will be useful to revisit what Polanyi himself says about the structure of comprehensive entities. Given that he explicitly develops this theory in the context of his unfolding epistemology, our outline of the metaphysics strictly implied by the ontological aspect of his theory of knowing will begin there. In *Meaning* Polanyi reminds us that

> the structure of tacit knowing...includes a joint pair of constituents. Subsidiaries exist as such by bearing on the focus to which we are attending from them. In other words, the functional structure of from-to knowing includes jointly a subsidiary "from" and a focal "to" (or "at"). But this pair is not linked together of its own accord. The relation of a subsidiary to a focus is formed by the act of a person who integrates one to the other. The from-to relation lasts only so long as a person, the knower, sustains this... (*M* 37)

The structure of a comprehensive entity is here seen as an integration of myriad subsidiary parts, some of which are unspecifiable at the time of beholding and some of which are epistemologically unspecifiable, full stop. Again, it ought to be emphasized that the permanent nature of *some* unspecifiability provides an in-built bulwark against the charge of a naïve one-to-one correspondence.

In the cited epistemological description, Polanyi notes that the from-to relation only holds so long as the knower sustains it, however, as I have alluded to above, this clearly cannot be the case for the metaphysical comprehensive entity – in some cases, at least. Imagine, for instance, that I am looking at a slender stalk of bamboo, gently swaying in the breeze. My epistemological grasp of this comprehensive entity lasts as long as I sustain it, but when I turn back to my work, the swaying stalk does not cease its swaying. Polanyi, however, never specifies with precision the integrative force that sustains the relations obtaining between subsidiary particulars making up an unobserved comprehensive entity. I will soon argue, however, that an expanded understanding of comprehension and meaning will help fill this gap.

Returning to the structure of comprehensive entities, Polanyi notes that even though in most cases there will be unspecifiable subsidiary details involved in the makeup of a gestalt, theoretically "the relationship of the particulars jointly forming a whole may be ineffable, even though all the particulars are explicitly specifiable" (*PK* 88). This is the other side of his teaching that "dismemberment of a comprehensive entity produces incomprehension of it and in this sense the entity is logically unspecifiable in terms of its particulars" (*SOM* 45). Polanyi is not a bundle theorist.

The whole is thus irreducible to its parts and is no mere mereological sum, even if we were to have unbridled perspicacity as to its subsidiary structure. This teaching leads Polanyi to develop his theory of dual control, whereby the properties of the subsidiary substructure provide boundary conditions or parameters within which the emergent comprehensive entity's functioning must play out even as the emergent whole harnesses and directs in telic fashion the more basic properties of its constitutive parts. Crucially, this telic harnessing cannot be reduced to or predicted on the basis of the properties of the subsidiaries, according to Polanyi. He notes that

> Since our knowledge of… a particular comprehensive – i.e. 'molar' – achievement…is not specifiable in 'molecular' terms, the function itself is not reducible to molecular particulars; it must be acknowledged therefore as a higher form of being, not determined by these particulars. We can reach this conclusion directly by recalling that the understanding of a whole appreciates the coherence of its subject matter and thus acknowledges the existence of a value that is absent from the constituent particulars. (*PK* 327)

And also that

> If each higher level is to control the boundary conditions left open by the operations of the next lower level, this implies that these boundary conditions are in fact left open by the operations going on at the lower level. In other words, no level can gain control over its own boundary conditions

> and hence cannot bring into existence a higher level, the operations of which would consist in controlling these boundary conditions. Thus the logical structure of the hierarchy implies that a higher level comes into existence only through a process not manifest in the lower level, a process which thus qualifies as an emergence. (*TD* 44–45)

Subsidiaries therefore exercise bottom-up control through the limits and potentialities of their properties while the comprehensive entity exercises top-down control through the operation of its irreducible telic harnessing and direction of those potentialities in ways similar to Robert Koons' staunch hylomorphism (2014). Thus arises what is often taken by Polanyi himself (as well as his interpreters) to be a recommendation of a hierarchical ontology, wherein lower subsidiaries are integrated into higher, irreducible comprehensive wholes, as is especially clear when Polanyi applies his theory in a critique of Neo-Darwinian evolutionary theory. It is easy to understand the widespread focus on ontological stratification in Polanyi's work, since he so often draws attention to the way that a "comprehensive entity and…its particulars…represent two distinct levels of reality [wherein] obtains…a peculiar logical relationship, derived from the distinction between subsidiary and focal awareness" (*SOM* 46).

Yet this emphasis on hierarchy and stratification can be problematic in several ways. First it tends to focus on a narrow set of examples that center around the emergence of higher level, immaterial principles that are irreducible to the physical components that make up the lion's share of the specifiable subsidiary details – typically living things and machines. These test cases involve an obvious leap in Eleatic potency, from the relatively inert to the active and perhaps even sentient. Such examples are important and worth discussing, however they are not representative of the full range of beings properly encompassed under the term comprehensive entity, and thus run the risk of leading us to assume that there is such a radical Eleatic leap inherent in *all* subsidiary-whole relationships, but this is not so.

Consider that subsidiary particulars, which we generally take as simples when we tacitly indwell them in focusing on the *gestalt* to which they contribute, are quickly recognized as comprehensive entities unto themselves once they become the objects of our focal gaze. Yet this means that the subsidiary-whole stratification would be more appropriately designated *mereological rather than ontological* since it is possible for comprehensive entity *A* (richer in terms of IFM-production) to function as a subsidiary particular in comprehensive entity *B* (which happens to be less powerful in Eleatic terms relative to *A*). To sum it up another way, participating as a subsidiary particular in a given comprehensive entity *does not* entail the metaphysical inferiority of the particular, though in some cases (arguably, living things and machines, for instance), subsidiaries will generally be metaphysically inferior in terms of IFM production. What *is* entailed by the subsidiary-whole relationship is *mereological integration entailed by ontological constitution.*[9]

Furthermore, in direct contrast to Marmodoro's No Distinction Doctrine, the participation of a given subsidiary detail in another comprehensive entity *does not exhaust the being of the subsidiary detail as a comprehensive entity unto itself.* A subsidiary element maintains its own identity as a substance, if you will, even as it is integrated into other comprehensive entities.[10]

Thus the common emphasis in Polanyi scholarship on stratification overlooks a side of Polanyi's metaphysics that has gone more or less unremarked: The mereological hierarchy of subsidiary particulars and comprehensive entities must logically unfold within a *flat ontology of comprehensive entities.*

Additionally, focus on living things and machines (the classes of things designated as comprehensive entities in the era of *Personal Knowledge*) can lead us to assume (wrongly, in my view) that all comprehensive

---

[9] This issue will be revisited. For now, consider the example of the only extant letter addressed to William Shakespeare. The letter was written by Richard Quiney (1598) and involved a request for Shakespeare's assistance in securing a loan. It is not clear whether Shakespeare received the letter or not. This letter (a comprehensive entity of minor Eleatic potency) includes as a subsidiary detail in its mereological structure its addressee, William Shakespeare, himself a comprehensive entity of much greater Eleatic standing. For details on the letter see https://shakespearedocumented.folger.edu/resource/document/only-surviving-letter-shakespeare-letter-richard-quiney-asking-shakespeares.

[10] Here the position I am urging on the mereological relations of subsidiaries to comprehensive entities to which they contribute (which I believe to be consonant with Polanyi's incomplete account) is akin to positions taken by Johnston (2006), Koons (2014), and Jaworski (2016) in hylomorphist debates over whether parts in a hylomorphic compound lose their own identity when they are "re-identified" (Marmodoro 2013, 15). Marmodoro argues that this re-identification does entail the loss of identity of the parts, (which she calls "fusion") (2021) while the others disagree.

entities will involve physical subsidiaries. Yet if we recall Polanyi's understanding of the marks of the real (and especially the IFM effect), then it becomes evident that the reality of comprehensive entities *qua* comprehensive entities is in no way determined by physics or tangibility for Polanyi. Minds, problems, tunes, abstract patterns, artistic trends, plotlines, amoebas, shards of glass, rabbits, and cobblestones are real comprehensive entities. Many comprehensive entities are thus physical, material beings, while many others are non-physical, immaterial beings. Yet all such comprehensive entities, be they physically tangible or no, share the same ontological structure. Furthermore, those entities that are physical are shot through with the immaterial insofar as they are *comprehensive* and *comprehensible* in a way reminiscent of substantial form and principles of unity in Aristotelian hylomorphism. In the following section we will examine the relationship of the material and the immaterial in a Polanyian metaphysics. For now, consider the following passage in which Polanyi muses on the reality of the immaterial mind:

> The mind is unsubstantial only in the sense in which a problem is unsubstantial. Indeed, a great mind is an inexhaustible and rewarding problem to the historian and literary scholar, and every person is of infinite concern to one who cares for him. *But neither problems nor minds should on this account be set far apart from other things. For an inanimate solid object, too, is known by understanding its particulars, from which we attend to it as an object*... With this in mind, we can say now also that man's skillful exercise of his body is a real entity that another person can know, and know only by comprehending it, and that *the comprehension of this real entity has the same structure as the entity which is its object*. (*TD* 31 emphasis added)

Again, we know all these things due to our indwelling of their particulars, but, given the ontological aspect, we must posit that these things are sustained in their own being by the metaphysical structure of which both they and we partake.

6 Metaphysics, Ontology, and Mereology

As we have just seen, Polanyi's epistemology, when run through the ontological aspect, requires that we recognize a mereological hierarchy of subsidiaries and comprehensive entities wherein subsidiaries are themselves comprehensive entities in their own right. In addition, comprehensive entities (whether functioning as subsidiaries or not) may be either immaterial, material, or a combination thereof. This mereological hierarchy emerges within what I have argued ought to be characterized as a of comprehensive entities, given that all of the existing participants partake of this basic structure and all such integrated entities are real in the same sense.

My emphasis on mereology will lead us to focus in more depth on the details of this parts-whole relation soon enough. As I have indicated above, Polanyi's most detailed engagement with mereology comes in the form of his teachings on dual control, which focus on comprehensive entities such as living beings and machines that primarily involve material components as the (specifiable) subsidiary particulars which give rise to an emergent hierarchical structure wherein higher levels arise from the lower, but then, having arisen, regulate the lower levels, along with the regulatory principles of the lower levels themselves (each level is thus subject to dual control – by its own rules and those of the level above it). Again, the idea is that the higher levels are dependent on the lower for their emergence, yet then exercise certain sorts of regulatory priority over the lower (though such priority too is bounded by the lower levels in certain respects). Some relatively higher levels (such as the chemical) are material while some even higher (e.g. meanings, set theory, formal causes, perhaps self-reflective consciousness) are not. Immaterial entities thus tend to be classed as higher-level entities.

Furthermore, "higher" vs. "lower" in this scheme often align with the another, non-mereological hierarchy that Polanyi affirms when he argues that "that the kind of comprehensive entities exemplified by skillful human performances are real things; as real as cobblestones and, in view of their far greater independence and power, much *more* real than cobblestones" (*TD* 33). This claim rests on his Eleatic criteria of the real (addressed above in Chap. 2), especially the IFM effect.[11]

My proposal is that this Eleatic hierarchy is metaphysical, rather than ontological. This is to say that all it takes for a comprehensive entity to exist is to be sustained as an integrated, substantive unit. To the extent that this is fulfilled, cobblestones, solar systems, skilled human performances, and minds are on an ontological par, since they all exist as integrated, substantive units. This is what it means to say that we are dealing with a flat ontology.

Yet Polanyi's stance that a skilled human performance is *more* real moves us from ontology (the study of beings as comprehensive entities) to metaphysics (in my parlance the study of the Eleatic potency of beings). The interpretation and development of Polanyi's thinking that I urge is thus to take the Eleatic Principle as applied to cobblestones as indicative of a metaphysical hierarchy measured in terms of richness of IFM production operating in conjunction with a flat ontology of comprehensive entities. Thus while all comprehensive entities have equal ontological status *qua* comprehensive entities, they do not have equal Eleatic richness and therefore can be hierarchically ordered metaphysically.[12]

This leaves us with a tripartite overall structure: An Eleatic metaphysical hierarchy tracked by richness of IFM production, a flat ontology of comprehensive entities, and a mereology that describes the subsidiary-Gestalt stratification of those comprehensive entities. This mereological hierarchy, however, *does not* entail an ontological hierarchy (since subsidiaries are comprehensive entities unto themselves) and in which subsidiaries are not necessarily lesser entities in metaphysical Eleatic terms. Again, this mereological stratification *may or may not* align with the metaphysical Eleatic hierarchy – that depends on the comprehensive entity in question and its subsidiary particulars.

But it is not enough to say simply that the metaphysical backdrop is an Eleatic hierarchy, for, as Dorothy Emmett (1945) argues, *qua* metaphysical it will also be analogical. In the following chapter we will return to

---

[11] See also Marmodoro 2021, pg. 66 for more on the Eleatic Principle.

[12] I am indebted to Andrew Grosso for helping me to see this more clearly (in email messages to the author, June 1 & 5, 2021).

Emmett's work on analogy and metaphysics and its influence on Polanyi. For now let us note in passing that our ability to track and compare the Eleatic action of IFM production between, say, a cobblestone and a skilled human performance such as Abraham Lincoln's Second Inaugural Address, relies on a metaphysics of analogical relations. *Analogy* here does not refer to metaphorical description, but rather foregrounds sameness-in-difference as a metaphysical fundamental by drawing our attention both to the recognizable continuity of structure as well as the differences in the way the structure is instantiated.

Similarly, the mereological relationships of subsidiary particulars to comprehensive entities are themselves identifiable as such analogically. In other words, the grains of sand that make up a sand castle; the physical components making up a living being; the various subsidiary elements of driving a nail with a hammer; plot, character, and setting as subsidiary details of a novel – these can all be identified as subsidiary components, but only analogically, since the entities to which they contribute are so varied despite the analogical and metaphysical unity of their shared mereological structure and the ontological parity of the parts and the whole in terms of their structure as comprehensive entities.

The tripartite metaphysical structure I am proposing here – an analogical and Eleatic metaphysical hierarchy, a flat ontology of comprehensive entities, and a mereological subsidiary-Gestalt hierarchy – most definitely goes beyond Polanyi's own work, but I would argue that it is a textually defensible development of his ideas.

7 Immateriality

Returning to the comparison of the reality of skilled performances and cobblestones, it is evident, when taken together with the affirmation of problems and minds as real entities that we are guided toward the metaphysical conclusion that Polanyi is not only a realist with respect to abstract objects, but a realist who affirms abstract objects as capable of ascending the highest, most privileged metaphysical peaks. Despite this, the precise status of the immaterial in Polanyi's ontological hierarchy has been subject to significant debate.

As interpreted by Walt Gulick (1999, 1999–2000), for instance, the lower levels provide the material substrate from which the immaterial emerges, with meaning eventually accruing to endow life with existential richness. Yet he generally wishes to save the language of "reality" for reference only to those entities which are entirely empirical or which exhibit a mixture of meaning and the empirical (by which he means those entities which are themselves human-constructed meanings, but which give rise to empirical components). Some other interpreters, such as Paski (2010, 2011), assert baldly that Polanyi does not affirm the existence of any substance over and above the material. S.R. Jha, contemplating with mild distaste the ontological application of Polanyi's epistemology observes that "if he were consistent, he would have started with an ontology of infinite mind" (1999–2000, 18). Charles Lowney seems potentially open to such a move, noting without any sign of disapproval that

> If pressed …Polanyi might…admit to higher-level organizing principles in some inanimate systems. Hints of this inclination are found towards the end of Personal Knowledge where Polanyi speaks of a "generalized field" at work in organizing and developing natural systems (PK, 398)…one might even come to agree with Peirce, who held no strict divide here but spoke of physical systems and material objects as mind 'hidebound with habits." (2013–2014, 24)

Meanwhile Cannon (1999–2000), Rolnick (1992–1993), Mullins (1999–2000), and Sanders (1999–2000) are all, like Lowney, quite bullish about the prospects of the immaterial in a Polanyian ontology, though they don't quibble much about where in the hierarchical order it might arise. Esther Meek's (2017) interest, meanwhile, tends to lie more in Polanyi's epistemological realism, though one suspects she tends toward metaphysical realism with regards to immaterial entities.

What nobody calls into question, however, is that according to Polanyi, order (at least of the sort assumed and theorized by physics and chemistry) is definitely present in the lower levels, and meaning (variously construed) is present at the very least in the higher levels. The question I want to ask is this: Does not this order *qua* order imply that some sort of meaning is already present at the lower levels? I believe so, since for order to be present and differentiable at all something like formal causation is necessary. Given that for Polanyi, "meaning" often implicitly refers in part to formal causation, the presence of order implies the presence of meaning in a consistent Polanyian metaphysics. Before I make this argument in more depth, however, it will be necessary to examine another class of immaterial entity often acknowledged by Polanyi, the category of telos.

8 Telos

In *The Study of Man* Polanyi urges us to

> Take a watch to pieces and examine, however carefully, its separate parts in turn…[we] will never come across the principles by which a watch keeps time. This may sound trivial, but is actually of decisive significance. For the study of inanimate objects constitutes the science of physics and chemistry and the study of machines forms the science of engineering, and we may now conclude, therefore, quite generally, that the subject-matter of engineering cannot be specified in terms of physics and chemistry. Let loose an army of physicists and chemists to analyse and describe in utmost detail an object which you want to identify as a machine, and you will find that their results can never tell you whether the object is a machine and if so, what purpose it serves and how." (47)

The point here is that the telos of the watch is not present within the material components of the machine, but is an immaterial purpose presupposed by the watch as a comprehensive telic entity. Indeed, the watch is literally unidentifiable and incomprehensible as a watch to a being who cannot, for whatever reason, comprehend the immaterial purpose entailed by the machine as a comprehensive entity. It would, in this situation, be interpreted and identified as a concatenation of metal lumps, of molecules of various sorts, or what have you. The telos of a machine is thus outlined in the operational principles that harness and regulate the lower, material levels of the machine, and works within the constraints determined by the lower level boundary conditions. As Polanyi puts it in *Personal Knowledge*, "a physical and chemical investigation cannot convey the understanding of a machine as expressed by its operational principles. In fact, it can say nothing at all about the way the machine works or ought to work" (*PK* 329). Telos and its attendant operational principles in machines, living things, and in comprehensive entities more generally is thus a central metaphysical concept in Polanyi's work. In *Meaning* Polanyi and Prosch state that

> It is true that the teleology rejected in our day is understood as an overriding cosmic purpose necessitating all the structures and occurrences in the universe in order to accomplish itself. This form of teleology is indeed a form of determinism—perhaps even a tighter form of determinism than is provided for by a materialistic, mechanistic atomism. However, since at least the time of Charles Sanders Peirce and William James a looser view of teleology has been offered to us—one that would make it possible for us to suppose that some sort of intelligible directional tendencies may be operative in the world without our having to suppose that they determine all things. Actually it is possible that even Plato did not suppose that his "Good" forced itself upon all things. As Whitehead has pointed out, Plato tells us that the Demiurge, looking toward the Good, "persuades" an essentially free matter to structure itself, to some extent, in imitation of the Forms. Plato appeared to Whitehead to have modeled the cosmos on a struggle to achieve the Good in the somewhat recalcitrant media of space and time and matter, a struggle well known to all souls with purposes and ends and aims. (162)

This "looser view of telos" is the one that is operative in the post-critical metaphysics I am attempting to work out here, and it is appropriate that Polanyi refers to the possibility that such a view might have even been held by Plato, since that is precisely the direction in which we are heading.

Before delving deeper into questions of telos, let us pause for a moment to consider Polanyi's description of the comprehensive entity that is a flame. He argues that

> …a simple gas flame contains all that is relevant. It represents a phenomenon of constant shape, fed by a steady inflow of combustible material and releasing a continuous flow of waste products and of the energy produced by combustion. Once a flame has been started, its shape and chemical

> composition can be varied without extinguishing it. To this extent, *its identity is not defined by its physical or chemical topography, but by the operational principles which sustain it. A particular collocation of atoms may accidentally fulfil the conditions for starting a flame, but this accident itself can be defined as such only by its bearing on the system of ordering principles which establishes the possibility of stable flames.* (*PK* 384, emphasis added)

Possibility is an important term here, since the potentiality that holds open the possibility of flames itself molds and channels the matter which ignites (recalcitrant matter in the hands of the demiurge?).

Indeed, when the centrality for Polanyi of telos is foregrounded in the context of his Copernican realism and a metaphysics of comprehensive entities, his phenomenological emphasis on discovery and its correlate, invention, rings significant Platonic overtones that emphasize the ways in which immaterial concepts guide, form, and limit material reality:

> Discovery, invention—these words have connotations which recall what I have said before about understanding as a search for a hidden reality. *One can discover only something that was already there, ready to be discovered. The invention of machines and the like does produce something that was not there before; but actually, it is only the knowledge of the invention that is new, its possibility was there before.* This is no mere play with words, nor is it meant to derogate from the status of discovery and invention as creative acts of the mind. *I am merely referring to the important fact that you cannot discover or invent anything unless you are convinced that it is there, ready to be found.* The recognition of this hidden presence is in fact half the battle: it means that you have hit on a real problem and are asking the right questions…This…is beautifully represented…in Michelangelo's unfinished St. Matthew…We see there…how the artist is about to cut away the stone from around the figure that he has perceived inside the marble block. (*SOM* 25, emphasis added)

In the observations made here, and especially in the italicized portions, Polanyi emphasizes the metaphysical importance of potentiality, given that he attributes to uninstantiated ideas (such as the figure perceived within the marble block) real ontological status. He describes how a machine's inventor makes a series of skilled achievements which delves into particular areas within the realm of pre-existing possibility and actualizes some heretofore untapped potential.

This vision is deeply Platonist in its affirmation of uninstantiated abstract objects. After all, on Polanyi's own account the machine already existed as an abstract object prior to its "invention" – it existed, that is to say, as an immaterial set of possible relations oriented toward some telos. The inventor, by means of the heuristics of discovery, mentally approaches that abstract object and then sets about actualizing it by making it concrete. Yet if this account is correct, it would be more apt to call an *inventor* a *discoverer*, a fact which Polanyi tacitly acknowledges in his use of "discovery" and "invention" as synonyms above (*SOM* 25).

Furthermore, he notes that "we identify a machine by understanding it technically; that is, *by a participation in its purpose and an endorsement of its operational principles*" (*PK* 330, emphasis added). The language of participation further amplifies the Platonic resonance.

Polanyi's examination of machines is illustrative of the structure of comprehensive entities in general, though it leaves out the autonomous nature of the telos of living things and is thus incomplete. Interestingly, however, Graham Harman's OOO, referred to above, often uses the term *machine* to refer to objects, the central category in his proposed flat ontology. His usage of *machine* is in many ways mappable onto the concept of *comprehensive entities*.

Before too long we will examine the ways in which OOO might helpfully inform a more robust Polanyian metaphysics. First, however, we must proceed from our examination of telos in Polanyi's thought to a related concept, that of meaning.

9 Meaning as Being: Formal and Final Causes

In the discussion above we saw that the operating principles that are crucial to Polanyi's notion of dual control are closely related to the telos of the comprehensive entity in question. Polanyi also often refers to the downward, telic control over subsidiary particulars exercised by comprehensive entities as their *meaning*. Meaning is thus bound up with telos, which is bound up with the telic harnessing of the properties and potentialities of subsidiary details achieved by operating principles (in essence very close to the Aristotelian concept of substantial form). We must also keep in mind that the subsidiary particulars which make up the comprehensive entity are themselves comprehensive entities in their own right. We have been focusing on the downward pressure exerted by such harnessing, but of course this power of top-down control reverberates outwards and upwards as well, as the comprehensive entity's telos is achieved by means of its harnessing capabilities. But what does it mean for an entity to have such harnessing capabilities? It is in essence to act as or to be a formal cause.[13]

On my reading, then, Polanyi's terminology ought to be developed and updated to explicitly include a recognition that the *meaning of an entity is a reference to its integrative structure which just is its power of acting as a formal cause over its subsidiary particulars. We might even go so far as to say that this meaningful integrative structuring of the subsidiary particulars implies a sort of indwelling on the part of the comprehensive entity itself.* On this reading, then, we are not the only ones who comprehend. The other entities are comprehensible because they are comprehensive insofar as they themselves comprehend their component parts.

This more capacious understanding of the *comprehension* that grounds comprehensive entities not only in the integrative epistemological comprehension of the beholder, but also the comprehensive activity by the telic harnessing of the entity as a formal cause makes more explicit Polanyi's understanding of reality as unified and rational, indivisible along lines of *res extensa* and *res cogitans.*

Furthermore, the meanings of comprehensive entities exercise a sort of metaphysical priority over their subsidiary components since

> The true nature of a thing composed of different levels of reality was found revealed by its most comprehensive feature, forming its uppermost level. This feature must be recognized first, before we can appreciate the subsid-

[13]Terrence Deacon, like Alicia Juarrero discussed below, also has a way of linking dynamic systems and their attractors to Aristotelian notions of Formal and Final Causes. See Deacon's, *Incomplete Nature* (2011).

> iary role played in respect of it by the particulars forming the lower levels. (*SOM* 71)

This interpretation of Polanyi's understanding of meaning as a sort of substantial form or principle of unity – both the being of the thing and an explanation of its integrative power over its particulars – clearly rings hylomorphic overtones, but is also similar to Graham Harman's post-Heideggerian metaphysical observation that

> If the being of the hammer withdraws from all of its relations, it is equally true that hammer-being is not a present-at-hand atom descended from the skies to breathe life into mere derivative relations. Instead, tool-being is a form, a kind of formal cause that acts as substance with respect to its surroundings, but which is born only as a relational composite of its internal elements. (Harman 2002, 170)

Or again,

> In other words, the bridge is not simply an innocent entity that is later hijacked by the as-structure so as to manifest itself in such and such a way. Rather, the tool-being of the bridge has already committed a hijacking in its own right, appropriating bolt, cable, trestle, and asphalt, devouring them into its own being. The bridge, like any tool-being, is already a system, a machine in which numerous components are arranged in some form or another. The object is a kind of formal cause. (Ibid., 259)

Harman's tone tends toward darker hues than does Polanyi's, who would be more likely to say that that the bridge has integrated bolt, cable, and the rest, rather than hijacked or devoured them, yet the underlying concept of telic harnessing is the same, and it can fairly be identified as formal causation, again, in rough line with the Aristotelian tradition of principles of unity or substantial forms.

In a related vein, Alicia Juarrero attempts to provide an account "of mereological causality that allows a rethinking of both formal cause and intrinsic teleology" (2013–2014, 9) by taking a dynamical systems approach in conversation with Polanyi's work. Taking autocatalytic reactions as a jumping off point, her work is fascinating and worth quoting at length. Using Polanyi's understanding of boundary conditions as a jumping-off-point, she notes that

> The two aspects of boundaries, the inner and the outer, correspond to Polanyi's insight concerning from/at epistemological perspectives, but the point [she tries] to make…is an ontological one: that the symmetry break occurs thanks to closure of context-sensitive feedback and that the radical emergent is the newly formed relational, hierarchically differentiated whole, which, in turn, is not other than the lower level in any sense of substance or efficient causality. From the complex dynamical systems perspective, then, mind and body are not "two different things" as Polanyi claims (KB, 238), but decoupled levels in a hierarchically differentiated dynamics. The question of whether whole-part causality violates the laws of conservation suddenly dissolves. What has changed is that mereological causality has been revitalized and higher levels of organization therefore given their ontological—and causal—due. (Ibid., 10)

Thus she welcomes

> …the reappearance of teleological language. Not only can we now understand the workings of formal cause: decoupled from energetic considerations, formal causes are the operation of second order context-sensitive constraints acting as a process of selection (regulation, modulation). Patterns, which are not other than relations, are far from epiphenomenal; they are indeed causally potent. So top- down causality understood in this manner, I believe, is all one needs to refute supervenience à la Kim. If causally effective, the higher level is real. And not only is it causally powerful; it exercises its top-down causality on its terms, that is, according to criteria determined at the higher level. Final causality is also thus compatible with real, ontological emergence: not only is the telos not pre-established or predetermined; it is also self-organized and thus radically emergent. As mentioned earlier, formal cause (now reinterpreted as the operations of second-order top-down constraints) selects for inclusion in or deletion from an autocalytic cycle those molecules that enhance and improve the integrity or cohesion of the higher level. Final causes (now reinterpreted as the metastable coherence of the higher level organization) are the self-organized telos in evidence in autopoiesis and other autonomous processes. (Ibid., 11)

Juarrero's analysis is lucid and helpful, though the narrow scope of her example (autocatalytic reactions) mean that the explicitness and clarity of her description of the functioning of formal causation or telic harnessing may be overstated, or at least inapplicable with such simplicity and perspicuity in many cases.

Nonetheless, Juarrero's semi-Polanyian reading of the causal and emergent structure of the comprehensive entity that is an autocatalytic reaction directs our attention in a structured way to the causal motion or directionality of Polanyi's conception of meaning as being. Meaning's downward action – its telic harnessing of subsidiaries – constitutes its being as formal cause; its telos – that is to say, its upward and outward teleological orientation which produces IFMs – constitutes its being as final cause. While Juarrero's willingness to countenance radically novel and emergent ontological final causes does not cohere well with the post-critical Platonism I am developing here, this can be surmounted by turning our attention once more to a more robust ontological view of potentiality as seen in our discussion of discovery and invention above, which renders a picture of the emergence of such formal causes not *ex nihilo*, but rather from the matrix or field of formal possibility which sculpts all possible telos, often referred to by Polanyi as *fields*. As Iain Hamilton Grant puts it, we are gesturing toward

> a phase space of the Idea in unlimited not-being, that is, the always-becoming, in which the Idea acts as the limit-attractor towards which becoming never ceases to become, the *auto* or absolute approximated but never realized in the generated particular, the whole caused by and causing productive and intelligent nature. (Grant 2006, 45)

Thus – to return to the theme of comprehensive entities – in an entity's meaning, which is to say its being, formal and final causes are united symbiotically in a logically harmonious way. This is so because the integrative power exercised by meaning as a formal cause that reorders and molds subsidiary elements of reality in accord with its forward-looking orientation toward its own telos is precisely what gives rise to a comprehensive entity's indefinite future manifestations, which are, in Polanyian terms, the prime determinant of the entity's reality. Some meanings are of course richer than others, as our discussion of the IFM effect under the heading of Polanyi's Copernican realism would indicate. A richer meaning – one that produces more and greater/richer indefinite future manifestations – resonates more resoundingly and sends out stronger ripples through the surrounding reality in which it participates, perhaps in the process giving rise to the emergence of further comprehensive entities. As Polanyi puts it, moving from our phenomenological experience of knowing through to its metaphysical implications and basis,

> By dwelling in a harmonious sequence of sounds, we acknowledge their joint meaning as a tune: a meaning they have in themselves, existentially. Up to a point our acknowledgment of the existence of a living individual proceeds on quite similar lines. We appraise in it a significant orderliness, which as such means something in itself. But a living individual is altogether different from any of the inanimate things, like tunes, words, poems, theories, cultures, to which we have ascribed meaning before this. Its meaning is different, perhaps richer, and above all, it has a centre. The focus of our comprehension is now something active, that grows, produces meaningful shapes, survives by the rational functioning of its organs; something that can behave and acquire knowledge, and at a human level, can even think and affirm its own convictions. (*PK* 344)

The metaphysical stratification of comprehensive entities which unfolds on the basis of the flat ontology determined by the subsidiary/whole mereological structure of beings is primarily one determined by richness of meaning, most easily tracked by its IFM production (or Eleatic potency).

In our earlier chapter on Polanyi's Copernican realism we saw that some interpreters such as Prosch and Gulick, have argued that meaning only enters the world picture through the rise of life, with its introduction of centers of achievement. There is no doubt that proponents of this Proschian approach have a number of texts by Polanyi where he says (or seems to say) that meaning enters the world picture with the rise of life. For instance in *PK* he states that the bacillus' "self-controlled shape and structure, and the physiological functions serving its survival, set up a centre of self-interest against the world-wide drift of meaningless happenings" (387). Or again, "while the first rise of living individuals overcame the meaninglessness of the universe by establishing in it centres of subjective interests, the rise of human thought in its turn overcame these subjective interests by its universal intent" (Ibid., 389).

In my view there are several things going on here. First, I do not think Polanyi had firmly worked out his concept of meaning at the time of *Personal Knowledge*. In these passages "meaning" seems to mean what we might phrase in a more mature Polanyian vocabulary as "especially rich meanings," "meanings that involve consciousness," "meanings that might give rise to or involve self-reflection," etc. As the term *comprehensive entity's* usage is broadened in Polanyi's more mature work, it seems to me that *meaning* takes on a broader cast as I have glossed above.

In a recent essay, Gulick has argued that

> …any of the items Polanyi lists as having existential meaning, certainly words, tunes, theories, have in themselves only a certain physical pattern that becomes meaningful through human construction and usage crystallized in cultural habits of thought and practice. In other words, the heart of the Polanyian Semiotic [Gulick is] proposing regards the proactive person as the creative agent indispensable to any comprehensive theory of meaning. Uninterpreted entities have, at best, potential meaning. Consequently, [Gulick] reject[s] Polanyi's definition of external "existential meaning" and reserve[s] the term for an aspect of human experience. (2022, 14–15)

This formulation maintains the general Proschian tendency outlined above, but it serves as a clarifying foil to the view I argue for here in its assertion that "uninterpreted entities have, at best, potential meaning" (Ibid.). On a post-critical Platonist account, the potential/actual binary is not an ontological distinction, but rather a way of tracking the modality of a given entity's ontological weight. To put the matter frankly, a potential meaning is a part of reality which exerts causal force as an attractor insofar as that potential meaning might be actualized. Emergence, then is not an unheralded, bursting-in-upon the ontological scene, but rather a coherent shift in modality from potentiality to actuality which takes place within the ontological scene.

Furthermore, according to the view of Polanyi's metaphysics I am arguing for here, the Proschian interpretation must be rejected primarily on post-critical metaphysical grounds. Within the context of Polanyi's project this more segmented approach cannot be correct since, aside from the fact that such a segmented view re-introduces the fact-value distinction that Polanyi sought to abolish (as well as a more rigid subjective/objective dichotomy than he was willing to countenance), it is also the case that such a view would theoretically deny to inanimate comprehensive entities the integrative power as formal causes that they clearly exhibit, even in an empirical sense (as seen in Juarerro's example of autocatalysis), and a "right" construal of reality would collapse into the sort of reductionistic, epiphenomenal thinking that Polanyi set out to combat.

This is the case since, bereft of the power of formal causation, all comprehensive entities would "really" be mereological sums-of-parts or bundles, with any "higher" or "emergent" properties being nothing but additional mereological bits thrown on top of the heap. A lump of granite would cease to be a lump of granite and would "merely" be the subatomic particles or field-points that make up the relevant subsidiary details and so on. This is in clear opposition to Polanyi's explicit insistence that comprehensive entities are identified by their highest principles. To parry this, proponents of the Proschian view would need to be able to point to some sort of *meaningless formal cause* (which I take to be a contradiction in terms) or cede the ground gained by Polanyi in his fight against reductionism, positivism, and anti-realism, retreating to one of those positions or some combination thereof.

If, then, reality is characterized by a flat ontology of comprehensive entities, as I have argued here, and if Polanyi's understanding of comprehensive entities is a trustworthy account (so far as it goes), as I take it to be, then meaning goes all the way down:

> For, as many astute thinkers from Socrates on have seen, the world cannot be thought of as ultimately meaningful unless the organization of its parts is meaningful, that is, unless there is some point to the way things are put together or, at least, to the direction in which they are developing. This would mean that we would have to attain a view of the world in which the universe, per se, is not "value-free." (*M* 161)

Thus far, then, we have moved by means of the ontological aspect of tacit knowing to an examination of the evolution of the concept of the comprehensive entity in Polanyi's corpus as examined by Phil Mullins, to arguments for a more robustly developed flat ontology of comprehensive entities around and within which the metaphysical and mereological hierarchies – wrongly identified as ontological hierarchies, as I have argued, and that have been the focus of most scholars of Polanyi – can play out. This metaphysical proposal involves an understanding of meaning as being, an understanding which cements and explains the connection between knowing and being that is central to Polanyi's thought and which connects him to many ancient thinkers as well. Furthermore, our analysis of meaning fleshed it out more robustly in terms of its downward telic harnessing of subsidiary particulars (formal causation) in the service of its telos which manifests itself indefinitely in an up and outward orientation (final causation). I believe that this analysis takes the metaphysical concepts that Polanyi worked with and develops them more rigorously in a direction which is harmonious with the overall tenor of his post-critical project. We have also begun to introduce hints as to how a sort of Platonism has been lurking in a non-systematic fashion below the surface in much of Polanyi's thinking.

Soon enough I hope to make good on my promise of proposing a speculative post-critical Platonist metaphysics which further situates the ontology of comprehensive entities that sets the stage for knowing. First, however, I wish to explore some important ways in which my proposed ontology of comprehensive entities can be further enriched by interaction with Object-Oriented Ontology.

10 OOO and the Dark Side of the Moon

First let us note that, assuming the connection of knowing and being (and vice versa) central to Polanyi's work and tacitly at play in OOO, it becomes apparent that Harman's take on tool-being and the quadruple object resonate quite well with Polanyi's approach. For example, Harman points out that.

> equipment functions by pushing us beyond itself, by vanishing in favor of the visible reality that it brings about. By disappearing from view in this way, it allows the ultimate reference to swallow all of its component forces into an invisible system or network lying silently beneath it, In this first sense, the "reference" of the tool is the encountered finality or terminus to which it anonymously contributes: the miniature bolts "refer" to the bridge, while the carpenter's tools tend to vanish in favor of the house as a whole. (Harman 2002, 25)

It is a relatively straightforward task to see how this is consonant with Polanyi's approach to comprehensive entities as detailed above. The subsidiary bolts refer or point to the comprehensive entity that is the bridge. The carpenter's tools, to the house. Equipment has, in Polanyian terms, a from/to structure. A tool-being is a comprehensive entity. In the context of OOO, however, the from/to structure of equipment explains a swathe of reality that Polanyi does not adequately address—even in his discussions of "cosmos" near the end of *Personal Knowledge*– the ongoing existence of the world itself as a context within which our epistemological moves unfold, for "equipment is forever in action, constructing in each moment the sustaining habitat where our explicit awareness is on the move" (Ibid., 18).

A further point of metaphysical contact is the agreement between Polanyi and OOO on the coincidence, or simultaneity, of meaning and being, a link which cements the connection between knowing and being. On Harman's account, "the being of the tool and its meaning are one and the same…the being of the tool is what that tool actually is, the impact it attains, its anonymous labor behind all that is present-at-hand" (Ibid., 26). He admonishes us to

> Note the following: (1) The tool's "meaning" is nothing other than the visible termination of its underground action. Just as the meaning of a signal-arrow is the region to which it alerts us by pointing, just as a word evokes its meaning by deflecting attention away from itself, so too do cable and pillars "mean" the bridge-system by vanishing into it. Here, the meaning of equipment is the final explicit reality that it serves to bring onto the stage. (2) In another respect, meaning is quite the opposite. For instance, we know that the phenomenological method aspires never to be stupefied by

> the simple presence of any phenomenon; rather, this method earns its living by gradually drawing out the countless hidden layers of categorial structures implicit in the merest appearance. In this way, it pursues the meaning of the phenomenon. This second, concealed kind of "meaning" is analogous to what [Harman] call[s] the tool-being of the entity, an act of primal effectiveness that eludes every possible view. (Ibid., 25)

While I have attempted above to develop Polanyi's theory of meaning in constructive directions, Polanyi's own treatment of meaning does not include such an unpacking of its metaphysical implications, but generally treats it as an epistemologically transparent structure once grasped by indwelling the relevant subsidiary particulars. This is perhaps because Polanyi was concerned, in the face of twentieth century atrocities, to affirm our ability to commit in good intellectual conscience to beliefs – such as moral beliefs – that could be given no non-circular, non-committal justification. He does this laudably by showing that there are no beliefs – including beliefs in scientific truths – that can be given such justification, and by providing a positive epistemological program for facing up to this without falling into relativism. In a sense, then, he shows that knowledge is always incomplete, always uncertain, and never fully articulable, yet his emphasis is always on the ways we get things right by following through with our intellectual (spiritual?) calling to commit epistemically with universal intent.

There remains, however, a correlate of Polanyi's teaching here that generally goes unremarked, which is a *de-emphasis* in Polanyi's work on what lies on the other, darker side of the moon, the tacit, the unknown territory never to be mapped. I believe that OOO, in its greater emphasis on regions of metaphysical obscurity has something to offer Polanyians here.

Harman uses the term *machine* to refer to tool-beings (or quadruple objects) as substantive entities made up of smaller parts. Distancing himself from a standard interpretation of Heidegger, he argues, as we have seen, that real objects (or comprehensive entities) are "a kind of formal cause" that appropriate and harness subsidiary details into their "own being" (Ibid., 259).

This may be of help in propelling us toward an answer to an important question facing a coherent Polanyian metaphysics: At the epistemological level, the knower, as the center of achievement, performs the integrative task of attending through the subsidiary particulars to the joint meaning to which they point. Yet what is the integrative force that binds comprehensive entities which go unencountered by rational observers? For Polanyi's link between knowing and being seems to presuppose a world of comprehensive entities independent of our noetic activity. The idea of the emergent comprehensive entity (or machine, in Harman's terms) as a formal cause, explored above, begins to make some headway here. My own analysis of meaning as encompassing both downward formal causation and the upward and outward movements[14] of final causation implied in the entity's telos moves us further toward an understanding of the stickiness of comprehensive entities *sans* third party sentient participation. Yet these moves gesture even further down the road, since comprehensive entities, as contingent yet possible arrangements of subsidiary particulars, point toward the logical necessity of Platonic Forms.

Harman's continued development of tool-being as machine is also helpful in that it shows clearly how the flat ontology that I have argued is implied by Polanyi's theory of comprehensive entities puts relations and events on the same plane as entities, since

> …the tool-being known as "bridge" is only a functional relation among numerous subsystems arranged in a certain effective way; it is more substantial than any context it enters, and more contextual than any substance that inhabits it. The way in which bolt and nail and trestle are mutually arranged, contacting one another as something specific, is ontologically no different from the way in which I myself confront the bridge. Instead of the notion of an entity as a hard, durable unit, but also instead of a theory telling us that only contexts are real and not objects, we arrive at a concept of the entity as a certain respectivity of parts, a machine or formal cause. But if every entity is already made up of a set of relations, the converse is also true: every set of relations is also an entity. (Ibid., 260)

In the context of our development of a theory of comprehensive entities I would add that though every set of relations is an entity, it may well

[14] The upward movement refers to telic harnessing while the outward movement refers to the comprehensive entity's IFM production.

be an entity of very little Eleatic potency. And while the final clause in the passage cited above does not necessarily follow from Harman's explicitly stated premises, the idea that machines, tool-beings, comprehensive entities, etc. are "always part and whole" at once (Ibid., 278) is, I think, well-demonstrated.

Harman notes that this might lead some to object that this scheme leads to an infinite regress of endless decomposition, yet this is only the case if one is thinking reductionistically. From a Polanyian (or OOO) perspective, there *is* a sort of potential infinite regress, but also an infinite progress toward the synthesis of new entities, a tightrope upon which we can walk back and forth.

This goes for all objects, all comprehensive entities whether they be new or old, constructed by humans or not, since such existents are not "self-evident, atomic finalit[ies]" but rather "function in numerous different equipmental ways, swept up into countless larger systems," (Ibid., 23) many of which run the course of their existence with no human contact or participation.

This brings us back to another side of the question that surfaced earlier – the dark side of the moon – whereby Polanyi's theories largely pass over the question of the integrative forces that bind comprehensive entities untouched by human epistemological activity – namely, the question of the metaphysical interaction between such untouched comprehensive entities, pieces of the world "that resist one another, that forever caress each other, or wage war with one another" (Ibid., 33). For OOO, as for Polanyi, the answer to such questions cannot plausibly be naturalistic in the sense of closure under physics, since "we still have no idea how physical relations (or any other kind) are possible in the first place" (Harman 2007, 189). Furthermore, OOO holds that "it is not human consciousness that distorts the reality of things, but relationality *per se*" (Ibid., 193). In other words the other side of the moon is dark not just to us, nor solely to recognizably sentient comprehensive entities. No beings (or at least no contingent beings) experience total transparency with other entities.

The upshot of this stance, if correct, is that when two nonhuman comprehensive entities interact, they do not fully come into contact with one another anymore than does a human knower who comprehends the meaning of a comprehensive entity, but who does not, in so doing, *fully* encounter or comprehend the object in its totality. Finitude intervenes here. When fire consumes cotton, to invoke one of Harman's favorite examples (borrowed from medieval Muslim philosophers) neither the fire nor the cotton fully encounters or experiences the other. This is easily misconstrued as implying a sort of panpsychism. While the view is certainly compatible with panpsychism – and Harman does not rule it out – he is careful to note that the "experience" referred to above is not necessarily a conscious experience (though it could be and surely is, for instance, when a rabbit eats a carrot), but is a matter of relation, specifically of contact.

This is directly applicable to my own understanding of both sentient and non-sentient comprehensive entities as capable of *comprehending* and thus *indwelling* their own subsidiary particulars, the harnessing power of formal causation entailed by the meaning of the comprehensive entity itself, as well as the concomitant power of final causation as defined in its telos.

Despite the indubitable power of comprehension and indwelling, we and other comprehensive entities do not fully encounter our subsidiary parts, nor do we fully encounter, experience, or understand other beings we interact with as our lives unfold in and out of concert with the beings around us. This is the darker, more mysterious side of the tacit dimension of knowing and being which Polanyi does not tend to emphasize in his justifiably optimistic epistemological orientation, but which must, I feel, be acknowledged in a more robustly developed post-critical metaphysics.

11 Chapter Conclusion

Before moving on to investigate in some detail the concept of fields in Polanyi's work, his views on universals, and then to a more explicit and in-depth argument for a Platonist understanding of his project it will be useful to recapitulate the metaphysical position staked out thus far.

The scheme I am proposing refuses to cede any territory whatsoever to reductionism and brooks no ad hoc interventions which would inject new ontological categories into the metaphysical system midstream. To reiterate, my argument is that reality is best described with a tripartite metaphysical structure. At the level of metaphysics, of being itself, reality is Eleatic (in the sense that a given being's metaphysical status is determined by virtue of its causal power to produce IFMs), analogical, and hierarchical.

Within this metaphysics, beings are best described by a flat ontology of comprehensive entities. All things that exist – material and immaterial, potential and actual – are comprehensive entities. An entity *qua* entity is comprehensive – it is not dependent on third party sentient noetic activity (i.e. a beholder) for its comprehensiveness. All beings have equal ontological status insofar as they are comprehensive entities (i.e. they are real), but such beings do not all have equal metaphysical richness, which again, is measured in Eleatic terms.

Meaning is being and entails the powers of formal and final causation. Thus all beings are comprehensive insofar as they exist as meanings which harness the properties of their subsidiary particulars through the power of formal causation. By means of the power of final causation these subsidiary properties are channeled toward the fulfillment of the entity's telos, resulting in the production of IFMs. Again, such IFM production is one way we can get a sense of a comprehensive entity's metaphysical richness.

The structural link between knowing and being is strengthened by this understanding of meaning. All beings, as comprehensive entities, are comprehensible because they themselves comprehend and indwell their own subsidiary particulars in order to achieve telic harnessing. The ontological aspect of tacit knowing is thus rendered more perspicuous by a more thorough unpacking of the metaphysical structure of being that gives rise to the structure of knowledge.

Within this flat ontology operates a mereological hierarchy that is seen in the comprehensive structure of entities themselves in terms of the relations between subsidiary particulars and the joint meanings to which they contribute. This mereological hierarchy is distinct from the metaphysical hierarchy that can be tracked in terms of the richness of the IFMs produced by the being in the surrounding reality. Thus we can say that reality is hierarchical, but ontology is flat.

Since this metaphysics does not countenance a fact/value distinction, the entire scope of reality is value-laden; meaning goes all the way down. This does not, however imply that all values are concretely instantiated by all things at all times (or, contra a weak Aristotelian view of the necessary instantiation of universals, at any time), since concrete or material instantiation is not an ontological requirement. Values can exist as uninstantiated abstract objects, just as, in Polanyi's example above, the possibility of a flame exists prior to its concrete instantiation as a physical flame.

References

Cannon, Dale. 1999–2000. "Some Aspects of Polanyi's Version of Realism." *Tradition and Discovery* 26/3: 51–61.

Deacon, Terrence. 2011. *Incomplete Nature: How Mind Emerged From Matter.* New York: W. Norton & Co.

Emmett, Dorothy. 1945. *The Nature of Metaphysical Thinking.* London: Macmillan & Co.

Fennell, Jon M. 2017. "Is Polanyi's Emergence Reductive?" *Appraisal* 11/2: 23–36.

Gallardo de la Torre, Francisco. 2003. "La Epistemología de Michael Polanyi: Una perspectiva realista de la ciencia." PhD diss., Pontificia Universitas Sanctae Crucis.

Grant, Iain Hamilton. 2006. *Philosophies of Nature After Schelling.* New York, NY: Continuum.

Grene, Marjorie. 1977. "Tacit Knowing: Grounds for a Revolution in Philosophy." *The Journal of the British Society for Phenomenology* 8/3: 164–171.

Gulick, Walter B. 1999–2000. "Beyond Epistemology to Realms of Meaning." *Tradition and Discovery* 26/3: 24–41.

———. 1999. "The Meaningful *and* the Real in Polanyian Perspective." *Polanyiana.* 8/1–2: http://www.kfki.hu/chemonet/polanyi/; http://www.ch.bme.hu/chemonet/polanyi/

———. 2022. "A Polanyian Semiotic." *Polanyiana* 31/2: 3–37.

Harman, Graham. 2002. *Tool-Being: Heidegger and the Metaphysics of Objects.* Chicago, Il: Open Court Books.

———. 2007. "On Vicarious Causation." *Collapse II*: 187–221.

Jaworski, William. 2016. *Structure and Metaphysics of Mind: How Hylomorphism Solves the Mind-Body Problem.* Oxford: Oxford University Press.

Jha, S.R. 1999–2000. "Polanyi's Problematic 'Man in Thought': the Tacit and the Real – an Exploration and a Critique." *Tradition and Discovery* 26/3: 15–23.

Johnston, Mark. 2006. "Hylomorphism." *The Journal of Philosophy* 103/12. Special Issue: Parts and Wholes: 652–698.
Juarrero, Alicia. 2013–2014. "Downward Causation: Polanyi and Prigogine." *Tradition and Discovery* 40/3: 4–15.
Koons, Robert C. 2014. "Staunch vs. Faint-Hearted Hylomorphism: Toward an Account of Composition." *Res Philosophica* 91/2: 151–177.
Lowney II, Charles W. 2013–2014. "From Epistemology to Ontology to Epistemontology." *Tradition and Discovery* 26/1: 16–29.
———. 2017. "Overcoming the Scientistic Imaginary." *Charles Taylor, Michael Polanyi, and the Critique of Modernity* ed. Charles W. Lowney II. Palgrave Macmillan. 143–168.
Marmodoro, Anna. 2013. "Aristotle's Hylomorphism without Reconditioning." *Philosophical Inquiry* 37/1–2: 5–22.
———. 2021. *Forms and Structure in Plato's Metaphysics.* Oxford: Oxford University Press.
Meek, Esther Lightcap. 2017. *Contact with Reality: Michael Polanyi's Realism and Why It Matters. Eugene, OR: Cascade Books.*
Mullins, Phil. 1999–2000. "The Real *As* Meaningful." *Tradition and Discovery* 26/3: 42–50.
———. 1997. "Polanyi's Participative Realism." *Polanyiana* 6/2: https://www.kfki.hu/~cheminfo/polanyi/9702/mullins.html
———. 2006–2007. "Comprehension and the 'Comprehensive Entity': Polanyi's Theory of Tacit Knowing and Its Metaphysical Implications." Tradition and Discovery 33/3: 26–43.
———. 2018. "Michael Polanyi on Machines as Comprehensive Entities." Unpublished Manuscript.
———. Forthcoming 2022. "The Journal *Humanitas* as an Incubator of Polanyi's Ideas." *Tradition and Discovery.*
Paski, Daniel. 2010. "Emergence and Reduction in the Philosophy of Polanyi (I)." *Appraisal* 8/2: 34–41.
———. 2011. "Emergence and Reduction in the Philosophy of Polanyi (II)." *Appraisal* 8/4: 29–43.
Polanyi, Michael. 1962 [1958]. *Personal Knowledge: Towards a Post-Critical Philosophy*. Chicago, IL: University of Chicago Press.
———. 1964 [1946]. *Science, Faith, and Society.* London: Oxford University Press.

———. 1964. "Lecture 1: The Metaphysical Reach of Science." *Duke Lectures.* Delivered at Duke University, February 10, 1964. http://www.polanyisociety.org/Duke/Duke1.pdf

———. 1966. *The Tacit Dimension.* Chicago, IL: University of Chicago Press.

———. 1967. "Science and Reality" *The British Journal for the Philosophy of Science* Vol. 18, No. 3: 177–196.

———. 2013 [1959]. *The Study of Man.* New York, NY: Routledge.

Polanyi, Michael and Prosch, Harry. 1975. *Meaning.* Chicago, IL: University of Chicago Press.

Quiney, Richard. 1598. "Letter from Richard Quiney to William Shakespeare." *Shakespeare Documented*: https://shakespearedocumented.folger.edu/resource/document/only-surviving-letter-shakespeare-letter-richard-quiney-asking-shakespeares.

Rolnick, Phil. 1992–1993. "Polanyi's Progress: Transcendence, Universality, and Teleology." *Tradition and Discovery* 19/2: 13–34.

Sanders, Andy F. 1999–2000. "Science, Religion, and Polanyi's Comprehensive Realism." Tradition and Discovery 26/3: 84–95.

Witkowski, Jan. 2008. "Stalin's War on Genetic Science." *Nature* 454: 577–579.

The Material and the Immaterial in a Post-Critical Platonist Metaphysics

1 An Orderly Innovating Principle

In the final section of *Personal Knowledge* Polanyi paints a captivating yet incomplete metaphysical picture of the universe's sweeping anthropogenetic movement towards sentient, rational beings such as ourselves, noting that while

> This entire evolutionary achievement can be localized within a circumscribed material system…*the process must have been directed by an orderly innovating principle*, the action of which could have only been released by the random effects of molecular agitations and photons coming from outside, and the operations of which could only have been sustained by a favourable environment…From a seed of submicroscopic living particles—and from inanimate beginnings lying beyond these—we see emerging a race of sentient, responsible and creative beings. *The spontaneous rise of such incomparably higher forms of being testifies directly to the operations of an orderly innovating principle.* (*PK* 386 emphasis added)

M. E. Turkis II, *The Metaphysics of Michael Polanyi*,
https://doi.org/10.1007/978-3-031-41211-0_5

Particularly important here is the concept of the orderly innovating principle. For Polanyi the achievement of rational life constitutes the latest in a series of revolutions:

> While the first rise of living individuals overcame the meaninglessness[1] of the universe by establishing in it centres of subjective interests, the rise of human thought in its turn overcame these subjective interests by its universal intent. The first revolution was incomplete, for a self-centred life ending in death has little meaning. The second revolution aspires to eternal meaning, but owing to the finitude of man's condition it too remains blatantly incomplete. Yet the precarious foothold gained by man in the realm of ideas lends sufficient meaning to his brief existence; the inherent stability of man seems to me adequately supported and certified by his submission to ideals which I believe to be universal. (*PK* 389)

These "revolutions" tap into the immaterial field of possibilities or potentiality touched on briefly in the previous chapter as well as to the vague but suggestive language of fields that Polanyi uses in *Personal Knowledge* as well as *The Tacit Dimension* to frame the forces of metaphysical integration in the universe. These revolutions are a collective equivalent of discovery or invention, a process of realization and actualization that unfold along the "gradient of deepening meaning" (*M* 177) which seems in Polanyi's texts to be bound up with his fields. Questions of just what Polanyi means by "field" along with the concept's proper role within a more robust post-critical Platonist metaphysics will thus be addressed early on in this chapter.

This unfolding noogenesis naturally leads Polanyi to an affirmation of transcendentals, as evidenced in the passage above by references to ideals held to be universal. He elsewhere speaks highly of "a traditional

[1] Once again, we are confronted with Polanyi's equivocal use of "meaning" which gives rise, as noted earlier, to debates over the status of the immaterial in Polanyi's ontology. It must be insisted at this point that "inanimate" does not entail meaninglessness and that, furthermore, it is not even clear, given current controversies in physics and philosophy, that matter *is* inanimate. Nevertheless, Polanyi clearly sees matter – those swathes of reality susceptible to physical and chemical analysis – as inanimate and often speaks as if this inanimate nature were equivalent to meaninglessness. This, however, as I have already argued, hangs in untenable tension with the overall aims of his post-critical project. We ought therefore to read such passages as making claims not about meaning *per se* but rather the relative richness of meanings in terms of IFM production.

devotion to the spiritual objectives, such as truth, justice, and beauty," (*M* 203) and as we have explored elsewhere, his heuristics of discovery revolve around Eros, our innate attraction to beauty. Such transcendentals are of course a core element of the Platonic characteristics of Polanyi's thought. In the tradition of Platonism transcendentals are held to be Forms very closely related to and bound up in the Good. Given that Polanyi, like many others, thought of Platonic Forms as a particular type of theory about universals, a thoughtful examination of Polanyi's own understanding of universals is in order as well. The exploration of Polanyi's approach to universals (which is to characterize them as the joint meaning of the non-reductive set of entities contributing to a class – in other words a comprehensive entity), along with our discussion of his fields, will naturally lead us to a more fully developed engagement with Plato.

I wish furthermore to argue that for Polanyi's famous hierarchical scheme[2] in which lower, material levels are harnessed towards the fulfillment and enactment of higher, immaterial levels to be workable, it is necessary that some sort of immaterial entities must logically precede the most basic material levels.

After all, Polanyi's own field-language often indicates that he sees immaterial entities at work ontologically in the earliest logical and temporal moments of the universe – consider, for example, a "field of potentialities which evoke action," a "field of new potential meanings," and a "fiel[d that] evoke[s] and guide[s]" (*TD* 89), or "an aspect of nature seeking realization in our minds" and "the spell of a potential discovery striving to emerge into actuality"[3] (*SFS* 19). We have just seen the way he argues that "the spontaneous rise of…higher forms of being testifies directly to the operations of an orderly innovating principle" (*PK* 387). Yet his discussions of this originary field are left quite vague, with the vast majority of his attention devoted to examining the ways in which lower and higher material levels relate when and after life has entered the

[2] It is important to recall that this hierarchy, taken to be ontological by Polanyi and most of his interpreters, but which I have argued is really a metaphysical question, has often been taken as the near-entirety of his metaphysics, yet I have already argued that a robust post-critical metaphysics will be tripartite in nature, involving metaphysical and mereological hierarchies and a flat ontology.

[3] In his 1964 introduction to *SFS* Polanyi expresses doubt about this particular phrase (6), nonetheless I see similar conceptual orientations in later works as well.

picture, the point, as we have seen, at which interpreters of Polanyi are more or less in agreement that he would countenance immaterial entities (including meaning) arising from the temporally prior material order through the tacit, integrative achievements of living things which constitute life's hazardous teleological striving.

My earlier discussion of meaning[4] as being and as entailing the powers of both formal and final causation should indicate, in my view, a plausible explanation for the way in which all of reality as real is thoroughly shot through with the immateriality of meaning and thus constitutes, at least in part, what Elizabeth Grosz calls "the framing conditions of materiality that cannot themselves be material" (2017, 5).

Thus, as Jon Fennell (2017) has argued, Polanyi's emergentism is plausibly reductive in Thomas Nagel's sense (but not reductionistic, i.e., not reducible to merely material parts) 2012),[5] though Polanyi himself does not clearly explain how meaning goes all the way down to the field of potentialities. In this chapter I go further on this front than did my earlier discussion of meaning by contending that the explanatory power of field theory in Polanyi's ontological history can be improved upon by invoking the language of Platonic Forms.[6] For the purposes of the arguments I present here, whether Forms be construed as separate from the immanent

[4] Is thinking of meaning as an immaterial entity an unwarranted reification? I think not, unless we have some good reason for believing that meanings do not exist but rather that our use of "meanings" is simply a useful fiction. Any arguments, however, proposed for such a belief will necessarily depend upon similar "reifications" of relevant "meanings." If one takes Polanyi's thought seriously, as I do, then no serious or comprehensive understanding of our epistemic and metaphysical situation is possible without being a realist with respect to meaning. Furthermore, it is clear that meanings are not material.

[5] "A reductive account will explain the mental character of complex organisms entirely in terms of the properties of their elementary constituents, and if…the mental cannot be reduced to the physical, this will mean that the elementary constituents of which we are composed are not merely physical." (Nagel 2012, 54)

[6] Kyle Takaki notes Platonic traces in the Polanyian metaphysical orientations of Esther Meek and Charles Lowney (2018, 19). There will be more than mere traces in the argument presented here. Furthermore, from the vantage point of knowing centers of achievement such as ourselves, I take the Forms I argue for to be, in Roy Bhaskar's terms (2008), real, actual, and empirical insofar as they form a part of the intransitive domain of reality, are actualized (i.e. instantiated) by means of other real entities' participation, and are a critical part of our sensible experience of the domains of the real and actual. Note, however, that as the argument unfolds I will distinguish universals (a cognitive tracking mechanism) from Forms (metaphysical grounds of particulars' natures).

particulars which participate in them or whether they be construed as present in those particulars is not important in a strict, logical sense.[7]

That said, I – following a significant number of interpreters of Plato's corpus – tend towards the view that Forms are not separate from particulars, at least not in a manner that would imply some sort of Two Worlds Theory (though they *are* separate in the way that a joint meaning is "separate" from its subsidiary particulars). In fact, a flat ontology of comprehensive entities within a hierarchical metaphysics provides the raw materials for a new Platonist argument against Two Worlds interpretations. This set of interpretive moves ought to make Forms feel less exotic and shadowy to those versed in Polanyi's work and who evince some squeamishness about overtly affirming the immaterial as substantively and reductively basic, perhaps due to a tacit acceptance of modernity's notion of a radical phenomenal-noumenal divide between the immanent and the transcendent.

An understanding of both universals and the nature of particulars as grounded in Forms, robust comprehensive entities that wield great Eleatic power through their power to shape reality, would put flesh on the bones of Polanyi's skeletal concept of a field of potentialities, and participation (construed as a relationship of appearance to reality rather than mystifying connections between two worlds) would provide a plausible explanation of how and why novel and contingent beings, living or otherwise, are teleological and meaningful. On this view emergence does not imply any new coming-into-being, but rather the actualization of already existing potentialities – concrete instances of immaterial rational relations immanent in the material entities welling up from the fundamental immaterial field of meaningful teleological potential. Yet positing such Forms would also allow us to address the more fundamental question of how reality is potentially differentiable and therefore intelligible in the first place through the establishment of a horizon of possible pattern, regularity, etc.

[7] On this account, it may be that the immaterial potential for meaning logically antedates the material as a horizon of possibility (if the Forms are indeed separate), or it could be that these two ontological elements exist simultaneously and inextricably at all logical moments (if there is no separation of Forms, or if separation does not imply a Two Worlds theory). A complete demonstration of which of these is more plausible lies beyond the scope of this project. The possibility of the material being logically prior to the immaterial, however, seems to me ruled out.

I will begin by examining questions of how Polanyi's fields ought to be construed as functioning and how Platonic Forms can help make sense of the language of fields and gradients. The following section will focus on Polanyi's approach to universals and the relationship of universals to Forms which will lead to an examination of Plato, Aristotle, and Polanyi on the question of concept acquisition. Here I will argue that despite explicitly stating that the problem of universals is one of induction, Polanyi's own epistemological insights into the tacit basis of induction and the concomitant impossibility of formulating an explicit procedure thereof open up space within his post-critical project for a more Platonist conception of such access. Finally, having thus prepared the way, I will lay out a more complete form of my proposed post-critical, Platonist metaphysics, building on the work already done in the previous chapter on comprehensive entities.

2 Fields of Potentiality

In the closing pages of *The Tacit Dimension* Polanyi declares that

> This part of the universe, in which man has arisen, seems to be filled with a field of potentialities which evoke action. The action thus evoked in inanimate matter is rather poor, perhaps quite meaningless. But dead matter, matter that is both lifeless and deathless, takes on meaning by originating living things. With them a hazard enters the hitherto unerring universe: a hazard of life and death.
>
> The field of new potential meanings was so rich that this enterprise, once started, swept on toward an infinite range of higher meanings, unceasingly pouring them into existence, for the better part of a billion years. (TD 90–91)

In this passage the field of potentialities seems clearly to pre-exist matter, life and hazardous meanings. The "enterprise" as a whole is imbued with the meaning latent in the "inanimate" matter, and the language surrounding the "entry" of hazard into the universe with the instantiation of life must be taken as rhetorically effective, yet logically loose, since the passage deems the advent of life a part of the enterprise traced straight back to the field of potentialities, and, after all, a potential danger is still a danger – even if not an immediate one. The overall picture here is of a "gradient of deepening meaning" (*M* 177), not of a sudden injection of meaning into an otherwise meaningless process.

Walt Gulick distinguishes Polanyi's use of field-language in *The Tacit Dimension* from its role in the final pages of *Personal Knowledge*, arguing that in this later phase "Polanyi deflates the power of fields," grounding them "in focused human inquiry" and allowing for them to be "envisioned in either epistemic or ontological terms" while limiting their function "to evoke and guide rather than control" (Gulick 2021, 42).

Earlier on in his analysis of Polanyi's field theory, Gulick provides a close reading and analysis of the role of fields in *Personal Knowledge*, where, on Gulick's reading at least, the fields play an ampler role in their various manifestations. He zeroes in on three senses of the word "field" that are of help in understanding its role in Polanyi's thought: "(1)…open area[s] of land…characterized by continuity…bounded by something different…(2)…disciplinary field[s]…(3)…force[s] that act at a distance" (Ibid., 35) as in physics.

In *Personal Knowledge*, Polanyi develops his field-language to explain the operations of the tacit component of knowledge in problem-solving situations, arguing for a

> generalization of the field concept in a strictly biological sense, purified of any *arrière-pensée* of physico-chemical equilibration. All the operations of the 'tacit component' (whether self-centred or seeking universality, whether conscious or unconscious) will be subsumed under this field conception. All mental unease that seeks appeasement of itself will be regarded as a line of force in such a field. *Just as mechanical forces are the gradients of a potential energy, so this field of forces would also be the gradient of a potentiality: a gradient arising from the proximity of a possible achievement.* Our sense of approaching the unknown solution of a problem, and the urge to pursue it, are manifestly responses to a gradient of potential achievement; and when we identify a morphogenetic field, we see in it in fact a set of events co-ordinated by a common gradient of achievement. (*PK* 398 emphasis added)

Polanyi notes that muscular co-ordination, the operation of the central nervous system after sustaining significant injuries, and the process of searching for lost memories "offer further instances of apparently unformalizable operations" which "again are evidence of fields of forces derived from various gradients of achievement. (Ibid.) Soon thereafter, however, Polanyi expands the use of fields analogically, referring to "generalized" and "cosmic fields," uses that are, in my estimation, very close indeed to the sense used in the passage from *The Tacit Dimension* cited above.

Gulick argues that the use of field language in Polanyi's corpus is generally confused and off the mark. One of his (Gulick's) particular concerns is to defend Neo-Darwinian theory (as augmented in recent years) from Polanyi's critique. The particulars of debates over the status of Neo-Darwinian theory are in the main irrelevant to my argument here. What is notable, however, is that Gulick's disagreements with Polanyi on this score, as with the role of fields, are traceable in my view to a metaphysical dispute over the ontological status of telos. Consider the contours of Polanyi's argument in the following passage:

> The evolutionary process takes place in the germ plasm, but it manifests itself in the novel organism which the germ plasm potentially embodies. It occurs in one place and manifests itself in another. Hence if, contemplating the process of anthropogenesis, we are clearly convinced—as I am—that this is in fact so, *we are driven to assume that the maturation of the germ plasm is guided by the potentialities that are open to it through its possible germination into new individuals. We are actually facing then the operations of a phylogenetic field guiding anthropogenic maturation along the gradients of phylogenetic achievement*—as clearly as the embryologist faces morphogenetic fields derived from the gradient of ontogenetic achievement. Nor can we fail then to note that at least in some cases we can experience such gradients internally. We know clearly the approach of a recollection for which we have been racking our memory, and will tend to compare this with an ontogenetic maturation which re-produces things already achieved before; while we know also how the search for entirely novel achievements is guided by intimations of their growing proximity, even as the possibility of unprecedented achievements guides the maturation of the germ plasm to ever higher evolutionary stages. (*PK* 400 emphasis added).

What Polanyi addresses here is not so much a question of the biological details of evolution as a question of what sort of metaphysical foundations must underlie evolution's unfolding if we are to make sense of our empirical observations. The novel organism prefigured in the potentiality of the germ plasm "occurs in one place and manifests itself in another" (Ibid.). We are dealing, then, according to our observations, with a sort of action at a distance, a phenomenon often explained scientifically by means of field theory as evidenced clearly in electro-magnetism, gravity, and various other topics such as non-locality in quantum theory, about which more anon.

The metaphysical difficulty Polanyi tackles here by means of field theory is to posit an explanation of how a novel life form, which in its very instantiation embodies numerous meaningful achievements and is only intelligible in terms of biological achievement, can arise by means of such action at a distance. Since to a significant degree, the thrust of Polanyi's theorization of achievement (in its biological scope) is to make biology's infamous kept mistress of teleology "honest," what we are dealing with here is a teleological field of potentiality whose reality would explain how it is possible for such achievement-oriented evolutionary action at a distance to be logically possible or intelligible at all.

The question, then, is really of the ontological status of telos, which is to say, given my analysis in the previous chapter, meaning. We have already seen, of course, that for Gulick meanings and telos are not fundamental realities, though he is happy to take such non-fundamental entities very seriously once emerged. The crux here is thus Gulick's moderate, yet still notable, empiricist skepticism regarding immaterial metaphysical entities that leads to a tacit denial of real ontological status to telos and potentiality.

Apart from more particular questions of evolutionary biology, Gulick does not find the original analogy between physical and problem-solving fields to be helpful or illuminating, and so, unsurprisingly, he finds the concepts of the generalized, cosmic field which evolves therefrom to be unconvincing as well. It is important to note again, however, that his reading proceeds through the lens of his understanding of the real and the meaningful as two separate spheres, a view which ultimately throws an impassable barrier in the way of any *real* analogical validity, since on this view analogy is relegated to the sphere of meaning which emerges from the real but cannot, in a strict logical sense, be causally related to the phenomena it purports to explain if it is grasped as a meaning *after* the phenomena in a temporal sense. This is because on Gulick's view meanings are in a sense purely epistemic and do not have the same ontological status as the real. Since Polanyi's use of fields is metaphysical and ontological from the get-go, unapologetically positing real entities that are not directly empirically detectable, it is doomed from the beginning on Gulick's terms.

As I have already indicated, there are good reasons both for rejecting any strict ontological distinction between the real and meaningful and also for believing that Polanyi would have rejected such a distinction himself, despite some equivocation in *Meaning*. In fact, as Phil Mullins (2022) has unearthed, Polanyi was influenced significantly by his work and friendship with Dorothy Emmett, a metaphysician who argued that analogy is itself the core of metaphysics in her 1945 book, *The Nature of Metaphysical Thinking*, almost assuredly read by Polanyi. Many passages in this fascinating work seem to quite clearly prefigure much of Polanyi's epistemology as developed in *Personal Knowledge* and beyond. To the extent that this is true, it may provide further arguments for the primacy of metaphysics in Polanyi's thought, though that is not a line of investigation I will pursue here.

Interestingly, Emmett notes that "the term field was…creeping into other studies, such as biology, and even psychology, though it is not clear how far those who so use it recognize that it is an analogical term drawn from electro-dynamics where it has a precise meaning" (Emmett 1945, 89). Given his career as a physical chemist and his probable study of this text, it is thus likely that Polanyi was, to some extent at least, explicitly conscious of his own superior knowledge of the precise meaning of "field" when he chose to employ it analogically. He was also likely aware of its use by other thinkers who did not have his scientific background.

Emmett's central argument in *The Nature of Metaphysical Thinking* is

> that metaphysics is an analogical way of thinking. That is to say, it takes concepts drawn from some form of experience or some relation within experience, and extends them either so as to "say something about the

> nature of reality," or so as to suggest a possible mode of coordinating other experiences of different types from that from which the concept was originally derived. (Ibid., 5)

For Emmett, all metaphysical thought is analogical[8] and it seeks thereby to unify various domains of experience and knowledge into a coherent world picture. She later specifies that

> A metaphysical theory proper contains analogical conceptions which not only symbolize certain structural relations, but also certain qualitative distinctions. Hence they are closely connected with the type of theory we have described in considering religion and theology; we have seen how these grow out of certain attitudes to the world having a certain qualitative character. Like the religious judgments, the basic metaphysical judgments are of the nature of total assertions; they are judgments of importance and significance which govern the development of a theory. If you ask from whence are these basic judgments derived, I should suggest that they are derived predominantly from some particular type of experience, e.g. intellectual, aesthetic or moral, which has seemed to provide a clue in terms of which a *Weltanschauung* or philosophical attitude could be developed. The theory must then be developed according to canons of consistency and comprehensiveness...How far this is successful depends on the driving power of the original experience, and on the metaphysician's power of sustained constructive thought. (Ibid., 194)

I quote Emmett at length here in order help to clarify Polanyi's view of the role of analogical thinking in metaphysics from the more limited role it is allowed to play within Gulick's metaphysical vision.

It is important to note, however, that Polanyi freely used analogical components not only when addressing metaphysics, but also in his work on physical chemistry, as can be seen in his article "On Simple Gas Reactions," co-authored with H. Eyring in 1931, where he develops with his co-author a potential surface theory by creating an analogical topographical map of potential transition states which they represent as a

[8] Though Emmett is not a Thomist, she engages at length with Thomistic arguments about the *analogia entis* both in her book (1945) and, more briefly, in an article (1941) on analogy in metaphysics.

"resonance mountain range" (1931, 430). The result is essentially a representation of the field of potentiality for such chemical reactions, facilitated by the free use of analogical, geographic language – "saddle," "mountain," and "valley" (Ibid., 432). *This* is a sample of the sort of experience that Emmett argues is at the heart of all metaphysical theorizing and which leads Polanyi to posit his metaphysical fields.

Furthermore, we should recall that Polanyi regularly spent time with some of the founding figures of quantum physics, engaging them in dialogue and debate. It is thus almost certain that when he proposes a metaphysical field theory, he is integrating his knowledge of both classical field theory (broadly speaking, electromagnetism and gravitation) and the evolving understandings of quantum field theory (combining classical field theory with relativity and quantum mechanics) with which he was conversant.

It will be useful to note briefly, therefore, that vector, scalar, and tensor fields are all employed in physics. A scalar field might be used, for instance, to track and represent the temperature distribution throughout a room by assigning a temperature value to every point in the room's topography. Electromagnetic theory postulates fields in which every point or node in a three dimensional space is characterized by a directional force vector, allowing us to track the gradient of the field and to thus understand how action-at-a-distance (for instance the movement of a metal needle in response to fluctuations in the electromagnetic field) is possible. Increasingly throughout Polanyi's lifetime, tensors were used by physicists working in field theory in order to handle more complex relationships between both vectors and scalars.

When Polanyi speaks of gradients of deepening meaning, then, he is not merely speaking poetically. He is thinking of a vector space within which one can potentially identify or assign the meaning value for each point in the space and thus navigate, follow, or track – on the basis of tacit integrations leading to explicit, focal knowledge – the progress toward deeper meanings.

In a differently-oriented series of pieces (2004–2005) on Polanyi's field theory (which includes an earlier version of Gulick's critique as well as a piece by Joseph A. Bracken) sociologist Richard W. Moodey muses that

> A literal field is spatial, a bounded area of land, often fenced-in. Inside the boundaries or fence, plants grow, animals graze, or people play. A morphogenetic field is not literally spatial, but a metaphorical conceptual space. Because of the meaning of the Greek word genesis, I imagine these primarily as growing fields, only secondarily as playing fields. Forms, rather than plants, grow in morphogenetic fields. Both growing fields and playing fields have intrinsic characteristics, as well as boundaries. These constrain and facilitate patterns of growth as well as patterns of play. For example, football fields and baseball fields constrain and facilitate different patterns of play. (Moodey 2004–2005, 27)

Moodey's exploration of the metaphorical contours is instructive and fruitful. It draws our attention to the mix of constraint and facilitation, characteristic of all creative activity, that emanates from the fields, and it freely engages in the sort of serious analogical activity constitutive of metaphysics.

Moodey also points out that in advocating contemporary complexity theory as a better explanation than fields, Gulick himself engages in analogical thinking and that therefore, to a certain extent, his (Gulick's) preferred terminology amounts to a distinction without a difference. Moodey notes that

> It seems…that Gulick affirms the notion of morphogenetic fields, but gives them another name…"multi-faceted systems." He presents "contemporary complexity theory" as representing "emergent, stable patterns" (forms) "developing spontaneously" (growing) in "multi-faceted systems" (morphogenetic fields). The operative word here is so small as to be easily overlooked – in. By saying that these stable patterns or forms develop in systems, Gulick is using the same kind of container metaphor as Polanyi… *As containers in which emergent patterns or forms grow, multi-faceted systems have to be imagined as embodying "preexisting principles," whether "mysterious" or not. Otherwise they would not be systems.* (Moodey 2004–2005, 27–28 emphasis added)

This is an important point, as it underscores the inescapability of analogical thinking and serves to thus undercut any *ad hoc* distinctions between meaning and being.

Nonetheless, while I disagree with Gulick's general rejection of the usefulness of Polanyi's field analogy (and the metaphysical limitation on analogy that I have briefly argued is implied by Gulick's stance on the real and the meaningful), I do agree with him that Polanyi's use of field theory is yet vague and underdeveloped. I depart from Gulick, however in my belief (in alignment in with Moodey) that the evocative language of telic fields is a theoretical component that can and should be more sufficiently developed as a part of a robust post-critical metaphysics.[9]

I will now go on to argue that Polanyi's field conception can be strengthened and made more perspicuous in two ways: First, a forthright acknowledgement of the ways in which the field theory is implicitly Platonist will show that the language of fields is a way of referencing the real ontological status of potentiality, explicable in terms of the flat ontology of comprehensive objects proposed in the last chapter. Second, this first move towards additional clarity will be further developed by an in-depth treatment of a sub-category of comprehensive entities, Forms, which I will argue are the "substance" constitutive of the fields.

3 The Implicit Platonism of Polanyi's Field Theory

In contemporary analytic debates over universals, the dividing line between the two most prominent camps of realists (that is to say, those who affirm the existence of universals over against nominalists and trope theorists) runs between the Platonist position, which holds that uninstantiated universals exist (and thus that the particular is logically derivative), and the Aristotelian position, which holds that universals cannot exist uninstantiated (and thus that particulars are fundamental and universals logically derivative). Furthermore, the norm in these debates is to assume that Forms and universals, if not rough synonyms, are rival theories competing for the same metaphysical prize (Armstrong 1989,

[9] Other interpreters of Polanyi have also weighed in on his field theory in various ways. For an excellent exploration of Polanyi's conception of fields and the ordering principle, see Jon Fennell (2017) Sect. 1. See also Joseph A. Bracken (2004–2005), mentioned above, who combines Polanyi's field theory with a Whiteheadian approach to social organization.

6) – hence the possibility of claiming that the Platonic position holds that uninstantiated *universals* exist rather than holding that uninstantiated *Forms* exist. If, as Lloyd Gerson (2005) argues (following ancient Neoplatonist precedent), Forms and universals are *not* rival theories but rather seek to explain different phenomena, then it may well be that uninstantiated Forms can and do exist while uninstantiated universals cannot. This indeed is the conclusion I come to, but the arguments will be developed over the course of several chapters. Suffice it to say, however, that this lands us in a terminological pickle, since it makes the term "universal" a rather slippery fish. I shall strive for clarity, yet care will also be needed on the part of the reader.

For clarity's sake, then, I will briefly lay out the distinction I, following Gerson, take to hold between universals and Forms, though arguments for this position will be unpacked further on. I understand universals to be the conceptual and cognitive mechanism by which we track the real natures of particular entities to the extent that we are able to fallibly grasp such natures. Forms, meanwhile, are the metaphysical ground of sameness-in-difference that explains how different particulars can, despite their differences, also exhibit the same natures (e.g. Plato and Aristotle are different entities, but are both humans, men, philosophers, Greeks, etc.). As the ground of the real natures of particulars, then, Forms are also the ground of the universals we use to track those natures. Since I am advancing a Platonist argument, I will use the term *Form* to refer to the metaphysical ground for real natures sketched above as consistently as possible, and *universal* (with a lower case *u*) to refer to the entities that emerge mentally as we track such natures. There will be moments, however, especially when dealing with Polanyi's own writings on these topics as well as when handling objections from others who take the two terms to be interchangeable as described above, when such usage would be cumbersome. In those passages I will use the term preferred by the objector, but will capitalize it. Thus *Universal* (with an uppercase U) will refer to the usage common in many discussions which takes metaphysical grounding of natures *and* conceptual tracking of such natures to be inseparably bundled together into one metaphysical entity.

We will investigate these issues in more depth anon. I mention these distinctions here because I mean to demonstrate that Polanyi's field-language implies a Platonist position on Forms as existing uninstantiated due to its vigorous metaphysical support for the real ontological status of the potential. Here, however, we encounter our first *caveat lector*, since many interpreters of Polanyi as well as philosophers more generally, especially those suffering from an allergy to Plato, would likely prefer the phrase "uninstantiated Universals," again assuming a degree of interchangeability which I here reject.

Polanyi argues at the end of *Personal Knowledge* that in order to account for the rise of life in the universe that leads, at least in part, to our own sentience

> two things must be assured: (1) Living beings must be possible, i.e. *there must exist rational principles*, the operation of which can sustain their carriers indefinitely; and (2) favourable conditions must arise for initiating these operations and sustaining them. In this sense I shall acknowledge that *the ordering principle which originated life is the potentiality of a stable open system; while the inanimate matter on which life feeds is merely a condition which sustains life, and the accidental configuration of matter from which life had started had merely released the operations of life.* And evolution, like life itself, will then be said to have been originated by the action of an ordering principle, an action released by random fluctuations and sustained by fortunate environmental conditions. (*PK* 383 emphasis added)

On the next page, Polanyi gives a smaller scale example of a potential open system with which we are already familiar (a flame) that is actualized by a contingent and accidental arrangement of matter. We will return to the example of the flame shortly. What bears emphasis in both these examples is the metaphysical status of the potentiality of stable, open systems of particular kinds (whether a flame or life itself). Polanyi explicitly states that "there must exist rational principles" (Ibid.). The potentiality of stable, open systems is a description of the bare minimum of rational principles which must exist in order to account for, or in other words to metaphysically ground, the rise of life (or flames).[10] The logical

---

[10] See Kastner et al. (2018) on how granting the potential full ontological status renders intelligible "the counter-intuitive features of quantum mechanics such as nonlocality, entanglement, and instantaneous collapse" (169).

possibility of a particular kind of stable, open system is, in essence, a Form, an entity which holds open at the level of potency the opportunity for the instantiation of a kind. Even the phrase "stable, open system" itself refers to a more general range of possibility – the instantiation of a multiplicity of such systems which despite their variety exhibit sameness-in-difference.

Given that Polanyi clearly sees the real potentiality of such systems as "embodied" in the rational principles (i.e. Forms) prior, logically and temporally, to the instantiation of such systems in matter, he affirms the possibility of uninstantiated Forms and is thus himself a Platonist in this respect.[11] This relatively simple point is a crucial one, and we will return to it often in what follows: Polanyi's analogical field theory affirms, both implicitly and explicitly, a Platonist position with respect to rational principles, which is to say Forms, which exist uninstantiated as a necessary precondition of reality as we experience it.

Furthermore, the rational principle explicitly affirmed by Polanyi above as real, the Form that embodies the potentiality of stable, open systems, is itself susceptible to further analysis in terms of the more basic Forms in which it participates – stability, openness, and systematicity, at the very least. That is to say that it is a comprehensive entity that consists of a set of meaningful relations among its subsidiary particulars which are harnessed by the entity's being as formal cause and oriented forward and beyond by its being as final cause, issuing in the IFMs it sends out as ripples in the surrounding reality. In the case addressed by Polanyi above, the IFMs include life as it has actually arisen. This prefigures my argument that the real, uninstantiated Forms acknowledged by Polanyi are the connective metaphysical tissue between his field theory and his unfinished metaphysics of comprehensive entities, which I have attempted to make more robust.

In response to my argument that Polanyi is a Platonist in his affirmation of the reality of Forms, an analytic Aristotelian realist[12] who takes

[11] Note that those who prefer at this point the language of analytic debates on Universals could substitute "Universal" for Form throughout this paragraph without changing the substance of the argument. The conclusion would then be that Polanyi is a Platonist given his affirmation of uninstantiated Universals.

[12] That is to say, an Anti-Platonist Aristotelian.

"Forms" to refer to uninstantiated Universals might object that Polanyi does no such thing, arguing that the foregoing passage is perfectly compatible with a weaker form of Aristotelian realism which merely requires that a Universal be instantiated at some time and place for it to gain permanent ontological status (as opposed to the stronger form which holds that the Universal only has ontological status simultaneous with its instantiation – in other words the Universal redness only exists when it is instantiated in actualized red things). So the weak Aristotelian realist might say that the Universals of flames and living systems exist atemporally since they are in fact instantiated at some point in the time-space continuum.

I am convinced, however, that this is not a defensible reading of Polanyi's own position, since it effectively silences the *open* nature of the stable system. Polanyi is trying to grapple here with the interplay of necessity *and* contingency, which are both obvious components of reality as we encounter it phenomenologically. Reading Polanyi as a Platonist allows his analysis to do the work he intended. It acknowledges that reality is not bounded by the concrete or material, that forces of necessity are at play (e.g. real, field-like landscapes of logical possibility, his rational principles) and also that within these field-like boundaries, reality is dynamic and open. Not all is predetermined from the start. There are possible worlds in which no life, no flames are instantiated. In other words, rational principles must exist necessarily, but also, for contingency to be real, openness must be a reality. But if the rational principle of a stable, open system were to depend on its future instantiation for its own existence, then this would vitiate the very openness of the system and throw us into a strict determinism. The maintenance, therefore, of this pairing of necessity and contingency militates for a Platonist understanding of Universals, more appropriately dubbed *Forms.*

Forms, of course, are, *qua* Forms, intelligible structures. Polanyi's fields are themselves intelligible analogical structures that help us to understand reality itself as potentially intelligible, through and through. Thomas Nagel has remarked that

> The *intelligibility* of the world is no accident. Mind, in this view, is doubly related to the natural order. Nature is such as to give rise to conscious

beings with minds; and it is such as to be comprehensible to such beings. Ultimately, therefore, such beings should be comprehensible to themselves. And these are fundamental features of the universe, not byproducts of contingent developments whose true explanation is given in terms that do not make reference to mind. (2012, 17 emphasis added)

Jon Fennell has argued compellingly that Polanyi's emergentism is reductive in Nagel's sense, concluding that

[Polanyi's] account is reductive because that which is responsible for the constitutive emergence – the fields that give rise to new operational principles (and hence life) – exist, potentially, from the beginning. They do not emerge historically. Yet… unlike Nagel's consciousness which exists side by side from the beginning with mere physical existence, it is not the case for Polanyi that the operational principles (especially those defining living things) are eternal. Just as with Nagel's teleology, a bias toward development exists from the outset. But this bias can and does historically give rise to the unprecedented. (2017, Sect. 2)

I agree with Fennell that Polanyi's emergentism is reductive (though, again, not reductionistic) however I would maintain that it is not logically possible for such operational principles and biases toward development to be both truly novel and unprecedented *and also* prefigured by the fields, since the prefiguration sets the precedent, as it were, at the level of potentiality, which I insist be accorded full ontological status. The solution is that such principles do exist *potentially* from the get go as Forms, while those possibilities that do happen to manifest, also real, operate at the level of actualization, as can be seen simply represented in Polanyi's example of the flame, whose

identity is not defined by its physical or chemical topography, but by the operational principles which sustain it. A particular collocation of atoms may accidentally fulfil the conditions for starting a flame, but this accident itself can be defined as such only by its bearing on the system of ordering principles which establishes the possibility of stable flames. (*PK* 384)

The both/and-ness of Polanyi's dynamic picture of reality as both stable and open, operating on the basis and within the bounds of necessarily existing rational principles - yet with real contingency and unpredictability within those bounds - are central to the post-critical nature of the Platonist metaphysics I am arguing for. They are also a largely unrecognized part of his project and thus a natural direction for its further development. This interplay is on full display when Polanyi introduces the concept of a heuristic field:

> We assume that the gradient of a discovery, measured by the nearness of discovery prompts the mind towards it…The assumption of a heuristic field explains now how it is possible that we acquire knowledge and believe that we can hold it, though we can do this only on evidence which cannot justify these acts by any acceptable strict rules. It suggests that we may do so because an innate affinity for making contact with reality moves our thoughts—under the guidance of useful clues and plausible rules—to increase ever further our hold on reality…The lines of force in a heuristic field should stand for an access to an opportunity, and for the obligation and the resolve to make good this opportunity, in spite of its inherent uncertainties. It is true that the assumption of such a field expresses more clearly than has been done before that our expectation to discover the truth is justified by our nature as living beings…But in order to express correctly this kinship of knowing and living, fields must be interpreted throughout biology in accordance with their finalistic character, as fields of opportunity and of a striving directed towards this opportunity. (*PK* 403)

The lines of force represent an opportunity which is rife with uncertainty toward which we are drawn by virtue of the sort of things we are (living, sentient, etc.). The kinship of knowing and living is a more specific sub-type of the kinship of knowing and being investigated earlier in our discussion of comprehensive entities and meaning as being. It bears mentioning, however, that Polanyi's field theory is closely connected with another Platonist element of his thought briefly touched upon in my chapter on the nature of the post-critical project, the existential power of Eros, which is to say on intellectual beauty as a propulsive force which moves us to engage more richly with reality in fiduciary, post-critical fashion, even in the context of the hard sciences.

Thus questions of the status and nature of Forms (or Universals), often taken to be dry, sterile and abstract, are seen to have rich existential import at the level of metaphysics and phenomenology, driving to the heart of "the most general features of reality" (Armstrong 1989, 6). Polanyi ends *Personal Knowledge* with a breathtaking overview of his fiduciary vision of the rise of sentient life which consciously pursues the transcendentals:

> Looking back…on the immensities of the past, we realize that all that we see there, throughout the universe, is shaped by what we now ultimately believe. We see primordial inanimate matter, the motions of which are determined—whether mechanically or statistically—by intrinsic fields of forces. We see its particles settling down into orderly configurations which our physical theories can trace back (however imperfectly) to the fundamental properties of inanimate matter. This universe is still dead, but it already has the capacity of coming to life. Can we see then all the works of the human mind invisibly inscribed already in the configuration of primeval incandescent gases? No, we cannot; for the capacity of coming to life is due to the power of a field to consolidate centres of first causes. Each such centre bears a possibility of achievement which, however limited, uncertain, and unspecifiable in its outcome, characterizes this centre as an essentially new and autonomous prime mover. The centres of individual beings are short lived, but the centres of the phylogenetic fields of which individuals are offshoots go on operating through millions of years; indeed, some of these may endure for ever—we cannot tell. But we do know that the phylogenetic centres which formed our own primeval ancestry have now produced…a life of the mind which claims to be guided by universal standards. By this act a prime cause emergent in time has directed itself at aims that are timeless… pursuing…ultimate liberation. We may envisage then a cosmic field which called forth all these centres by offering them a short-lived, limited, hazardous opportunity for making some progress of their own towards an unthinkable consummation. And that is also, I believe, how a Christian is placed when worshipping God. (Ibid., 404)

There is great poetic power in this passage. Polanyi envisages the unfolding of reality as the dynamic interplay of various sorts of fields that allow for achievements in the erotic pursuit of richer meanings. He characterizes, as is typical for him, the universe prior to the rise of life as "dead." This is of course trivially true, but it is also metaphorically in tension with his characterization of the universe as containing or embodying rational principles which, once activated, seek liberation. This is already a meaningful universe which is pulsing with potential for richer meanings. Thus the principle of Eros is central to the rise of life.

In *Personal Knowledge*, Polanyi had still not yet generalized his concept of comprehensive entities beyond living things and machines, and he perhaps never saw that to be logically consistent they must be generalized into a flat ontology of all beings, as I have proposed. Nonetheless, reading this passage through the lens of such an ontology, in which all entities *mean* insofar as they *exist* (which entails that they *comprehend* by means of the telic harnessing of their subsidiary particulars and meaningfully move toward the embodiment of their telos) renders an even more fertile picture, in which the link between knowing and being does not flow in just one direction, primarily serving to metaphysically justify our epistemic powers, but rather spills over and makes palpable the ways that epistemic powers are themselves metaphysical and are shot through the entirety of the cosmos. Is this panpsychism? Some will say yes. I am myself unsure, yet at the very least it is a compelling invitation enter into Buberian I-Thou relations with wider and wider swathes of reality, an invitation which is surely consonant with the overtly religious finale and the mystical overtones of "unthinkable consummation[s]" (Ibid.).

The language of unthinkable consummations is evocative and dramatic, in part because it bristles with unresolved tension. For if I am correct in arguing that Polanyi's project entails a universe that is meaningful all the way down (despite his linguistic equivocations on this point), then logically speaking, a consummation *is* thinkable if it is possible at all, since if meaning is being, then to be, even potentially, is to be intelligible and thus thinkable. I prefer therefore to interpret this unthinkability as operating within the current scope of *actualized* reality (at any time t in the unfolding of the cosmos). That is to say, as finite knowers, a complete consummation is never *actually* thinkable, due to our own limitations within space and time, but it is *potentially* thinkable, and it is within our powers to grasp it in bits and parts. Such a consummation is thus a potent relation of Forms (which we track by means of universals) that functions as a point of erotic attraction creating intelligible lines of force within the heuristic and morphological fields that Polanyi theorizes. The phrase "unthinkable consummation" gains its poetic force insofar as it contains, though barely, this existential tension between the twin ontological realities, the potential and the actual, united by the common metaphysical structure of the comprehensive entity.

4 Universals

Having surveyed the scope and dramatic, erotic sweep of Polanyi's field theory that affirms the real ontological status of both the potential and the actual in its Platonist affirmation of Forms (i.e. the possibility of uninstantiated Universals), it is time to turn our attention to Polanyi's explicit teaching on Universals. For the time being, I shall use the term *Universal* rather than *Form*, since this is Polanyi's own terminology and his texts make it quite clear that he takes Universals to refer to the same metaphysical issues Plato's Forms purport to handle (needless to say, Polanyi does not follow the capitalization scheme I've established here, and I will not change his capitalization when quoting his works, or the works of others). As we finish our survey of Polanyi's explicit texts on Universals as well as our subsequent contextualization of his position thereon within my more robustly developed theory of comprehensive entities, I will return to a more strict usage of the Platonic language of *Forms*.

Polanyi is explicitly committed to the view that Universals "are the joint meaning of things forming a class" (KB 170). He also explicitly affirms the reality of Universals, declaring that "a joint focal meaning...dwelt in as [a] subsidiary clu[e] to such a meaning...points to yet unrevealed joint properties of its members" (KB 52). Clearly, if we can indwell an entity, that entity exists. Since the IFM effect (see Meek 2017), is one of his tests of the ontological status of an entity, it is clear that Polanyi is a realist with respect to Universals.

Indeed, he explicitly rejects the "cloak of nominalist respectability" when he accuses nominalists of avoiding "the question [of] how the same term can apply to a series of indeterminately variable particulars" (*PK*

113). He claims that such a dodge arises from nominalist thinkers' "abhorrence of [the] metaphysical alternatives," alternatives that he is explicitly willing to countenance "by accrediting the speaker's sense of fitness for judging that his words express the reality he seeks to express" (Ibid.). In *Meaning* Polanyi and Prosch offer the following anti-nominalist gloss on Universals:

> Yet the speculations of two and a half millennia, striving to eliminate the indeterminacy involved in subsuming a presumed instance of a class under that class, seem to have been misguided. This indeterminacy is irreducible, but its comprehension is safely controlled by tacit integration. Tacit knowing commonly integrates groups of particulars into their joint meaning. Members of a class such as a species, a family, or a language—or members of any other group properly denoted by a single universal term—possess a joint focal meaning when dwelt in as subsidiary clues to such a meaning, even though their focus is almost empty in contrast to a focal object of perception. Moreover, the meaning of a class is an aspect of reality, for it points to yet unrevealed joint properties of its members. (*M* 52)

This passage elucidates the way that Universals are at play in our epistemic tracking activity as well as reiterates the definition of Universals in *Knowing and Being* (52, 170) as joint focal meanings of a set of things forming a class. Considering this definition in light of my arguments for a flat ontology of comprehensive entities in which meaning is being, several things become clear that allow us to provide a more robust metaphysical theory of Polanyian Universals (i.e. Forms).

First, Universals ought to be construed as comprehensive entities whose subsidiary details are the particulars which participate in the Universal and the relations obtaining among them. On my flat ontological view this follows trivially. If all real entities are comprehensive, and Universals are real entities (as Polanyi and I both explicitly affirm), then Universals are comprehensive entities. There is also a textual argument to be made that even if Polanyi never explicitly refers to Universals as comprehensive entities, he *does* explicitly refer to them as joint meanings, which is one of his most common descriptions or definitions of the term comprehensive entity.

Second, as comprehensive entities, Universals exercise the same formal and final causal powers discussed earlier. The participation of subsidiary red particulars – balls, crayons, etc. – in the comprehensive entity *redness* is logically possible only because of the real *potential* for concrete redness held out by the *uninstantiated* Universal of redness. This is an explanatory description of what I have called the power of telic harnessing characteristic of comprehensive entities as meanings. There is, in a certain sense, a drawing of the particulars toward the Universal, a comprehending of the particulars by the Universal, which can also be aptly described using the language of participation bequeathed us by the Platonic tradition. There is also the forward-oriented potential for indefinite future manifestations (IFMs) of novel red things which constitutes its power as a final cause.

Third, this understanding sheds light on the way we are able to identify and track redness in various concrete particulars by means of our indwelling of those particulars in the familiar ways described by Polanyi's epistemology of tacit knowing. We are now able to see how, as Plato (1997) notes in *Phaedo*, we access the universal by means of the particular. By indwelling a red particular, we are indwelling the Universal redness itself insofar as the particular *is a subsidiary component of the comprehensive entity* Redness.

Fourth, this scheme also helps us conceptualize the way Universals are both transcendent entities in their own right and also immanent in the subsidiary participative particulars. When taken together with what I take to be Polanyi's correct Platonist affirmation of the possibility of uninstantiated Universals (implied by his field theory), a theory of Universals as comprehensive entities provides the structural foundation of a powerful Platonic argument against Two Worlds interpretations of Platonism, an argument I will develop presently after an in-depth examination of Polanyi, Plato, and Aristotle's views on how we access Universals. Briefly stated, this argument holds that the structure of Universals (or Forms) as comprehensive entities with both potential and actualized instantiations as subsidiary details militates against seeing the immateriality of the Forms as a radically separate Platonic heaven somehow divorced from the sensible world. It rather pushes us towards seeing the sensible world as shot through with the immateriality and transcendence of the Forms, which provide the deep framing conditions necessary for the intelligibility of the material elements of the unified world we experience, which is both transcendent and immanent, ideal and material.

Fifth, we are also able to see relatively clearly the way in which Universals as comprehensive entities are both universal (in that their IFMs are novel subsidiary participants)[13] *and* particular (in terms of their existence as the joint meaning of their class of subsidiaries). This allows us to freely incorporate Aristotelian insight into our post-critical Platonic theory while bypassing a common understanding of Aristotle's classification in *The Categories* of Universals as secondary substances on the basis of his putative assumption that only particulars can be fully substantive.[14]

Sixth, understanding Universals as comprehensive entities provides clues to their metaphysical structure which combines a closed, definitive nature (redness) with an open-ended, creative potentiality (IFMs consisting of novel red things). In this sense we can think of Universals as stable, open systems, a concept that figured importantly in our discussion of fields above. As Polanyi notes,

---

[13] Note that this specific type of IFM production is *not* the case for all comprehensive entities. For example, I (Martin Turkis) am a comprehensive entity capable of a wide range of IFMs. Those IFMs, however, will not be further instantiations of myself, but rather things like this text, my children (who are not simply little Martin Turkises based on me as a paradigm, but are themselves unique comprehensive entities), footprints I leave in the sand, etc.

[14] In *Categories* Part V we read:

> Substance, in the truest and primary and most definite sense of the word, is that which is neither predicable of a subject nor present in a subject; for instance, the individual man or horse. But in a secondary sense those things are called substances within which, as species, the primary substances are included; also those which, as genera, include the species. For instance, the individual man is included in the species 'man', and the genus to which the species belongs is 'animal'; these, therefore - that is to say, the species 'man' and the genus 'animal, - are termed secondary substances.

However, Adrian Pabst (2012), taking a Platonist approach, notes that

> The opposition of the one and the many seems perennial and perhaps even indelible. However, to oppose them is simply to assert that they represent contrary principles and that as such they are incompatible, producing conflict and violence. This opposition is based upon the undemonstrated and unwarranted assumption that substance is prior to relation and that individuality is absolute. (location 250)

> In this changing world, our anticipatory powers have always to deal with a somewhat unprecedented situation, and they can do so in general only by undergoing some measure of adaptation. More particularly: since every occasion on which a word is used is in some degree different from every previous occasion, we should expect that the meaning of a word will be modified in some degree on every such occasion. For example, since no owl is exactly like any other, to say 'This is an owl', a statement which ostensibly says something about the bird in front of us, also says something new about the term 'owl', that is, about owls in general. (*PK* 110)

Thus this conception of Universals provides a clear refutation of nominalist arguments such as Nietzsche's leaf example in "On Truth and Lies in a Nonmoral Sense" (1979). For Polanyi, Universals with full ontological status are a necessity if we are to understand our own tacit, integrative epistemic powers, as is clear in his argument that a Laplacean mind would be unable to recognize comprehensive entities purely on the basis of its physicalist pointillism (*PK* 54).[15] In essence, his realism with respect to Universals is part and parcel of his rejection of physicalism and reductionism as metaphysically inadequate. In embracing this metaphysical trajectory with regard to Universals, he recognizes that he is, in important respects, throwing in his lot with the ancients, though he does this without sacrificing a proper understanding of contemporary science. In a passage which carries overtones of Plato's *Cratylus*, with its concern for proper names as carrying metaphysical weight, Polanyi argues that

[15] ...let me assume once more that we possess the powers of a Laplacean Mind capable of establishing the complete atomic map of any object and also of computing this map for all future points of time. Let us imagine that these powers are to be applied to a living frog. In order to do this we must first know frogs and be able to identify a particular one, and we must also be able to distinguish living frogs from dead ones. Without the anterior knowledge of these comprehensive features the Laplacean enquiry could not be started. But even so, the Laplacean information that we gained would mean little, unless we could discern from it such further comprehensive features as the existence of different organs and their respective functions, including not only their machine-like operations but also such regulative or 'organismic' processes as maturation and regeneration. But could we derive these further features from our Laplacean prediction of atomic configurations? We could do so only by discovering the corresponding physiological shapes and patterns within these configurations; and for this we should have to rely on the same faculty for identifying comprehensive entities by which the physiologist normally establishes such features from the observation of living animals. (*PK* 54)

> In evaluating the differences between the archaic and the modern approaches, we have to maintain that the archaic mind is better in many ways. It is right in experiencing names as part of a named person and an image as part of its subject; for a name is not a name, nor an image an image, except as a subsidiary to the focal center on which it bears. And such is the nature of all meaningful relations. Admittedly, the archaic mind tends to exaggerate this coherence to the point of absurdity, but it is closer to the truth than the modern view, which has no place for the quality and depth of these coherences nor, therefore, for the full extent of the subsidiaries that are necessary to their composition. This difference becomes essential in the observation of those comprehensive entities that can be observed only by indwelling. The archaic mind recognizes indwelling as the proper means of understanding living things. (*M* 144)

5 Forms

Earlier I argued that Polanyi's analogical field theory entails a Platonist position with respect to Forms insofar as it postulates rational principles as a logical precondition for reality. In the section immediately preceding this one I used the nomenclature of "Universals," the term that Polanyi himself used. Having fleshed out Polanyi's thinking on Universals in the context of a metaphysics of comprehensive entities, we are left with a conception that is Platonist in the robust and substantive standing it gives to such Universals. Having set forth my reasons for thinking Polanyi a Platonist on this front, I will therefore now refer to such Universals as *Forms*.

I have noted earlier that universals are arguably distinct from Forms, as Gerson (2005) argues. Some other interpreters of Plato, however do take Forms to be u/Universals (Ademollo 2013; Sedley 2013). I wish at this point simply to point out that this debate within Platonism is not a sticking point for my project going forward since on my post-critical reading Forms – comprehensive entities which in their metaphysical grounding capacity give rise to both particular concrete instantiations and the cognitive universals we use to track them as potential subsidiary details – *do* contain as mereological parts the related universals/concepts used to track them. There is a sense, then, in which Forms *are* universals/concepts, but only partially. And it is important to recognize that such universals can also be examined as comprehensive entities in their own right.

My position is thus in harmony both with Gerson (2005) and the Neoplatonists he follows, as well as with Platonists such as Ademollo, Fine, or Sedley who do think of Forms as Universals since such thinkers do not understand Forms to be *mere* universals. In the parlance of this latter group of intepreters Forms go beyond being mere universals insofar as they are, to the extent that substance matters to Platonists, a substance, which is to say that they are real entities "that exist 'by themselves'" (d'Hoine 2011, 279) Substance, however, is a much more central concern for Aristotle, whose concept has to a very great extent set the parameters for future discussion of the topic.[16] Pieter d'Hoine reminds us of some key distinctions between Platonic and Aristotelian substances:

> Aristotelian substance is a category of being, albeit the most fundamental one. The things belonging to this category are subjects of accidental predication. Platonic substance, however, is not a category. It is simply a mode of being common to all the Forms *qua* Form, in opposition to the derivative existence proper to sensible particulars. For the Platonist, Forms alone can be called substances in a strict sense, for they are the only ontological items that are really self-subsistent...and self-identical...(d'Hoine 2011, 290)

In *Categories* Aristotle holds that universals or kinds are secondary substances, emphasizing the primacy of the particular and individual, which he categorizes as primary substances.[17] The closer the entity is to the level of the instantiated, concrete individual, the more substantive it is – thus

[16] Insofar as Anti-Platonist Aristotelianism has exercised significant influence in unfolding philosophical debates.

[17] "To subordinate reciprocal relation to individual substance is to elevate the self-identity of individuals over above the commonality of being in which all things seem to share. And to conceive individuality in this manner is to ignore particular specificities that are only manifest in the positioning of individuals in mutual relations with each other and their shared source of being." (Pabst 2012, location 252)

a genus is less substantive than a species, which is less substantive than a sub-species, etc.[18]

In the context of my argument for a post-critical Platonism developed on the basis of Polanyi's insights, my use of the term *Forms* can be taken as an endorsement of the following:

1. A commitment to the necessity of rational principles as a logical precondition for the stable, open systems we might encounter in actualized reality (i.e. Forms).[19]
2. A broadly conceived understanding of substance tied to the Eleatic Principle/IFM production and the concept of comprehensive entities discussed at length above and which includes such immaterial or abstract entities as Forms.

As noted by d'Hoine above, most Platonists have traditionally considered instantiated sensible particulars to be lesser, non-substantive entities, and this has gone along with a lack of a developed theory of matter within the Platonic corpus. We might thus paint with a broad brush and say that while Aristotelians want to ontologically demote Universals to the lesser rank of secondary substance, traditional Platonists tend to do roughly the same with sensible particulars. The post-critical Platonism I advocate here, however, differs from the Anti-Platonist Aristotelian position in that it maintains the substance of the Forms while it is also distinct from traditional Platonism in that it does not demote particulars and sensibles to a lower ontological or metaphysical status. I thus throw in my lot with Gail Fine when she notes that

> Though Aristotle goes too far in barring [Platonic Forms] from the ranks of substances, he can reasonably press a weaker claim: that [Aristotelian Substances] belong in the ranks as well…[thus] we should not accept as legitimate the presupposition of Plato's and Aristotle's question 'What are the substances?' – namely, that there is some one sort of entity that has privileged status, or alone satisfies all the criteria for substance. Different

[18] Justin Broackes (2006) provides a helpful gloss on Aristotle's substance scheme as developed in *Categories*. See especially pages 134–135.

[19] Or, in common analytic parlance, uninstantiated Universals.

> sorts of entities – both universals and particulars – are indispensable features of...the world. (Fine 1983, 44)

This is, in effect, the thrust of a flat ontology of comprehensive entities, which is inclusive of both of the universal and particular and thinks them together. Eric Salem, commenting on Graham Harman's approach to a flat ontology sums up the approach I advocate succinctly, noting that it aims to avoid the ontic biases inherent in erroneous conceptions of substance,

> All objects, natural and artificial, share the same formal structure, and in terms of this structure, a clock is as good as a cat: just as a cat is something over and above its parts, so too a clock is something over and above its gears and springs. In both cases the being of object-substances proves to be form. (Salem 2017, 233 citing Harman)

Adrian Pabst (2012) argues that

> ancient and modern conceptions of being in terms of individual substance fail to account for the irreducible, ontological relations that bind immanent, finite beings to each other...*Instead of some abstract 'individuality' that begs the question, [Plato] suggests that a thing's unique form is both existentially and essentially its metaphysical positioning in relation to other things. In turn, the relational ordering of all things suggests a priority of relation over substance, which intimates a first principle and final end that is itself* (Pabst 2012, xxiv emphasis added)

Such an emphasis on relation as constitutive of substantiality and not logically subsequent is but another way to speak of the structure of comprehensive entities, a relational network of subsidiary particulars teleologically harnessed by a joint comprehensive meaning, producing IFMs. Furthermore, on Pabst's account, this attention to the integral foundationality of relation is part of the Platonic inheritance, since "Plato links the constitutive relation between the Good and all the other...forms to the participatory relation among eternal forms as well as between them and temporal things" thus emphasizing "the priority of the relational `giving' of presence over participation wherein the Good infuses all things with goodness" (Ibid. location 265).

In important respects, post-critical Platonism resembles the Neoplatonist synthesis of Plato and Aristotle in its balancing of these concerns (relationality and substance) and in its effort to blend the insights of each approach. This, as has been noted earlier, is thematically appropriate to a development of Polanyi's project since he looks to Augustine, himself deeply informed by the Neoplatonist synthesis, as a model for his post-critical revolution.

In concluding this section, let us note that both Forms and sensibles are substantive insofar as they all partake of the metaphysical structure of the comprehensive entity, whether instantiated or not. And Forms as comprehensive entities are both universal and particular.

In my usage the capitalization of Form does not refer to a privileged ontological status as substance, therefore, but rather to the logical priority Forms hold in their relation to sensible particulars,[20] since Forms can exist as Forms uninstantiated, while sensible particulars, *even in potency*, are logically preceded by the Forms which they instantiate through their participation in the stable, open systems which are delineated by the Forms.

Forms likewise exercise the same logical priority over the universal concepts by which we track the natures of particular existents which Aristotle holds that we acquire through the process of abstraction (a topic we will examine more fully in later chapters). This logical priority operates at the level of the metaphysical hierarchy (within which the flat ontology unfolds), where it metaphysically grounds its derivative IFMs. It is also, in effect, yet another reference to the real ontological status of potentiality as referenced by Polanyi's language of fields, a topic to which we will now turn once more, this time equipped with a more perspicuous theory of Forms as comprehensive entities which guide and constrain the development of sensible reality.

---

[20] Lowney emphasizes this logical priority of Polanyi's universals, Frege's concepts, or Plato's forms and sees it as a reason to grant them ontological status. See Lowney "The Tacit in Frege" (2008, 21–26).

6 Form and Field

No doubt some interpreters of Polanyi, many of whom operate in a generally Aristotelian fashion with regard to substance (whether consciously or not),[21] will not yet be fully convinced by my arguments in favor of a flat ontology of comprehensive entities which affirms the existence of Platonic Forms and posits that Polanyi's predilection for talk of originary fields is closely bound up with the existence of such Forms. Such thinkers will likely prefer to remain within the conceptual scheme (common in Polanyi studies) of a hierarchical universe in which real immaterial entities emerge from the lower, physical levels, all conceived within the bounds of ontology. Such readers may feel that they've been taken on a bit of a wild, speculative metaphysical ride, and they may feel that such an approach is out of accord with Polanyi's own.

What I wish to do at this point is to tease out a bit more the ontology of Polanyi's originary fields in terms of Forms. In so doing I hope to connect the dots in a slightly different way that may, perhaps, be more convincing to such readers. Whether or not it succeeds on this front, I aim to explain more explicitly how Platonic Forms as comprehensive entities can explain satisfactorily the functioning of Polanyi's field-conception.

In order to proceed, we will return once more to the question of what, more or less precisely, is present for Polanyi in these fields that would allow for the rise of operational principles and their instantiations in the wide range of modes (physical, mechanical, living, and on through the artistic and mythical) acknowledged by Polanyi's metaphysics. We will also begin this process with a different set of uncontroversially Polanyian concepts.

Which concepts might those be? Well, considering the importance of boundary conditions in Polanyi's discussions of operational principles, boundary or limit, itself by itself, seems a clear example of such a Polanyian staple.

[21] I will not go into depth here on this topic, but it will be addressed more thoroughly later.

Let me develop this by taking another, perhaps rather obvious example. The orderliness[22] we recognize in the lower physico-chemical levels presupposes as a metaphysical backdrop the existence of concepts of pattern and regularity (e.g. boundary or limit, etc.) and their possible instantiation. Such a backdrop is, however, necessarily immaterial and non-spatiotemporal, even if such concepts or kinds are construed as immanent in the participating particulars, since the largeness in Simias and the largeness in Cebes are not identical, yet are, as it were, two recognizable appearances of the same trait (sameness-in-difference).

We have already dwelt at length on Polanyi's embrace of the reality of Forms, so I will not belabor the point further. What I want to look more closely at now are those most basic, elemental Forms (pattern itself, for instance, or limit itself) which are necessary subsidiaries of the very concept of *order* as embraced by such Polanyian thinkers and which therefore function as a part of the necessary tacit baseline that allows for any further integrative leap into the gestalt as is required by Polanyi's epistemology, which is affirmed by such interpreters.

For thinkers such as Fennell (2013) the Universals embraced by Polanyi are real,[23] but are emergent, arising only *after* achievement-oriented beings have engaged in integrative inductive acts of knowing and have thus cognized the universal. Gulick (1999) would take a similar position although he might insist that the Universals are not *real* but only *meaningful.* Yet logically, concepts such as Pattern or Limit cannot exist *merely* as the result of tacit integrations on the part of such beings since they (Pattern and Limit) are logically presupposed *by those very integrations.*

This raises questions about concept-formation to which we will return later. For now, let us inquire as to the nature of such basic, elemental concepts – concepts which must be logically present and accessible to account *not only* for the ability of living beings to tacitly integrate subsidiary clues *but also* for the very differentiability and order (i.e. orderliness) within the purely material elements (assuming for the moment that a

[22] We must recall here that interpreters such as Gulick and Prosch assume that order does not imply meaning.

[23] Technically, on Gulick's account such Universals would be "meaningful" but not "real" given their status on his view as products of human or animal cognition.

purely material order is a logical possibility). These concepts would set the horizon for intelligibility itself and necessarily logically precede the potential for achievement-oriented living beings. They are the metaphysical framing conditions necessary for the grounding of sensible reality.

Candidates for this role would not include entities dependent upon tacit integrations for their ontological status (though living beings' epistemological route to focal knowledge of such basic, grounding concepts may necessarily lie in tacit integration, as both Polanyi and Plato affirm), since such integrations, and the affordances they provide, are only intelligible in the light of logically prior conceptual possibilities, whether tacit or explicit. A subsidiary detail is, after all, in addition to being a constituent part of the tacit integration, an entity unto itself distinct from the other subsidiary details involved in the integration. Thus the very possibility of integration *qua* integration must presuppose Division itself, Boundary itself, etc., in order for the integrator to be able to fuse those boundaries in the creative act.[24]

The argument that I am tracing here is that if one accepts the Polanyian notion that tacit integration is the backbone of knowing, then one ought to be committed to a metaphysics that would provide the possibility for knowing beings by positing the logical necessity of the existence of at the very least those immaterial concepts that are necessary for the *orderliness* of the supposedly pure, lower levels of the stratified conception of the universe affirmed in much Polanyi scholarship. Furthermore, if there was a time *t* at which the lower, physical levels did not exist, but *could* exist potentially under some set of circumstances, whether necessary or contingent, then those concepts (such as Pattern or Limit) existed as real, uninstantiated universal entities – Forms –which provided the context in which, say, the Big Bang might occur, eventually giving rise to life, us, etc. Entities that are so fundamental as to delineate the very contours of intelligible reality clearly warrant the label substance if anything does,

[24] Note that the logical dependence of integration on subsidiary details' participation in Forms of Boundary and Division holds regardless of whether one is considering a particular integration from a first- or third-person perspective since the claim here is that such dependence is a logical precondition of all possible integration.

and this intuition is at least as powerful as Aristotle's purported stipulation that all substances are particulars.[25]

Interpreters such as Richard Moodey [26] insist that properties such as spatial dimensions or boundaries are not substances. I, however, would argue that Dimensionality, itself in itself, *is* substantive insofar as it is a necessary precondition for such a compound to be possible, let alone recognizable, at all, since this is an Eleatic sign of its existence. On my view the dimensions of some particular sensible would best be construed as the relation of the sensible comprehensive entity to the Form of Dimensionality as mediated by its (the sensible's) indwelling of or participation in the Form.

I would further point out that whatever else may form a part of Polanyi's fields of potentiality capable of evoking action, such Primordial Forms, as I will now call them, certainly must play an important role since they provide the logical gradient within which any potential operating or guiding principles for other ontological levels must take shape. By "Primordial Forms" I mean those Forms which *must necessarily* be instantiated for physical reality to exist and be intelligible at all. Thus Dimensionality itself in itself, Pattern, Limit, etc. are Primordial Forms. There are no possible worlds in which physical reality can get off the ground, so to speak, without instantiating these in various ways. This is not the case for all Forms, even those referring to natural kinds such as Tree, Snail, or Great White Shark; there are possible worlds that entail a physical reality in which these Forms are not instantiated. Therefore

[25] See Gail Fine's "Plato and Aristotle on Form and Substance (1983) for a detailed look at Aristotle's critique of Plato. Note also that precisely what Aristotle's position is on whether Universals can be substance or not is a matter of serious debate among scholars of Aristotle (see *Stanford Encyclopedia of Philosophy's* entry "Aristotle's *Metaphysics*," Sect. 10 for a helpful summary of the debate). In most debates between Platonists and Anti-Platonist Aristotelians over Universals, however, the standard Aristotelian position is taken to be that Universals exist as secondary substance, as outlined in *Categories* Part V:

> Substance, in the truest and primary and most definite sense of the word, is…the individual man or horse. But in a secondary sense those things are called substances within which, as species, the primary substances are included; also those which, as genera, include the species.

For examples of this see the section on neo-Aristotelianism in Chap. 3 above, Armstrong (1989), or Fine (1983).

[26] Richard Moodey, email message to the author, November 29, 2019.

while Tree is a Form, it is not a Primordial Form. We will develop this distinction further later on. For now, let us return to the fields.

Philip Clayton misses the point entirely when he quips (quoting *PK*), "but we now know that it is not necessary to postulate that the growth of the embryo 'is controlled by the gradient of potential shapes,' and we don't need to postulate a 'field' that guides this development…Stem cell research shows that the cell nucleus contains the core information necessary for the cell's development" (2002–2003, 13).

Let us consider a particular embryo, once again, an instantiated compound of form and matter. And let us but vary the use of the term "shapes" slightly in order to make an important point: Could the cell nucleus contain core information directing the cell to form a square circle? If the answer to this question is no (and it is), then why? Primordial Forms are here hemming in the possibilities, and this is therefore a gradient of potential shapes which limits the horizon of *all possible* material development, not just embryonic cell development. Yet, once we recognize this, it ought to be quite easy to see that even the logical possibility of the development of cells (i.e. temporally antedating the rise of life in space-time) is guided by the logical constraints exercised by immaterial components of reality which partially determine the possibilities for reality's emerging material components. It is important to note that such logical constraints are not merely "conceptual" in the cognitive sense – they are at least as real as the empirically detectable cells themselves, which are, after all, only accessible to us through the medium of concepts. They thus deserve a place in our ontology and our metaphysics. And if the example of a square circle seems too flippant, then consider that what Polanyi and Plato are after is the necessary metaphysical background for cells, nuclei, and information to be intelligible (and therefore possible) at all.

Some may worry that such a "shadowy ordering principle seems more like a first cousin to Platonic Ideas than a reliable foundation for scientific knowledge," (Gulick 2002–2003, 35) yet it is difficult to see how any intelligible entity with a material component (i.e. a hylomorphic compound) could come to exist or to be known without just such a backdrop, and this is precisely why Polanyi consistently recurs to the language of the evoking field. And at any rate, as David Bentley Hart correctly notes:

> We need not assume that prior to the complex unity of a living system some extrinsic design existed within its material substrate like a kind of algorithm programmed by an intelligent designer,[27] *but we can't doubt that everything that enters into the structure of a living system is already constituted by those rational causal relations that allow discrete purposive systems to arise.* (2018, emphasis added)

This is precisely the point Polanyi makes when he insists on the logical necessity of rational principles for the development of stable, open systems such as living things or flames.

If it is still unclear how Polanyi's fields (conceived of as populated by Platonic Forms) nourish, shape, and constrain the development of physical reality, consider James Franklin's semi-Platonist plea to his fellow neo-Aristotelians:

> ...as the physical world expands, mathematical necessities constrain what is possible. Although "constrain" cannot be read strictly causally, as if theorems can literally exert force on physical things to behave in certain ways, the "constraints" have visible consequences or "make a difference" in the world of causes, in such a way as to call in question the minimalist's earlier contention that uninstantiated properties have no causal effect. For example, the mathematical theorem that there is no continuous function from the circle to the real numbers that is increasing all the way round the circle means that it is impossible to build a spiral staircase that goes up all the way round and comes back to where it started – at any point, it may go up, but not at all points... *If "causality is the mark of being", power over the causal realm should also be a mark of being.* The close link between uninstantiated properties and what can happen in physical reality again suggests a maximalist, semi-Platonist but not fully Platonist view of the uninstantiated. (Franklin 2015, 32–33 emphasis added)

The "full Platonism" that Franklin rejects is what he calls "the 'extreme' or 'full-blooded' Platonism normally discussed in the philosophy of mathematics, according to which Platonic forms are strictly considered

[27] Ironically such algorithmic encoding, *sans* explicit reference to an Intelligent Designer, is quite close to the spirit of Clayton's comment above which gestures toward a sort of genetic homunculus that calls the shots (2002–2003).

denizens of an acausal and abstract world outside of space and time" (Ibid. 26–27). He notes, however, that "Plato's own Platonism may have been more nuanced" (Ibid.) Indeed it was, as is the post-critical Platonism I am advocating here, which does not subscribe to any such Two Worlds Theory. Addressing once more his fellow Aristotelians who are more leery of Plato than he, Franklin notes that

> Minimalist Aristotelians such as Armstrong argue that admitting uninstantiated universals in any way at all would be excessively Platonist, as acknowledging a realm of Forms beyond the real world, ungrounded in any true reality. They must say, then, that lengths greater than the diameter of the universe or uninstantiated shades of blue are mere possibilities. The difficulty for that suggestion is that "merely" possible lengths appear themselves to stand in ratios to each other, in ways correctly described by mathematics, and an uninstantiated shade of blue appears to lie between two determinate instantiated ones. The "mere" possibilities thus themselves form a Platonic-like world of forms, complex in structure, the truths of which have no apparent truthmaker. Our knowledge of ratios, such as that three times a length lies between two and four times that length, applies to lengths beyond the diameter of the universe. *Those truths stand ready to be, so to speak, clothed in reality if the universe expands.* (Ibid., 38 emphasis added)

Franklin's argument is thus that for an Aristotelian view to make sense of the relationship of uninstantiated mathematical objects (Universals in his lexicon, Forms in mine) to physical reality, it must move in a Platonist direction. In making this argument he clearly lays out the way in which, contra Clayton, the postulation of something like fields and gradients cannot be shrugged off, since even the presumably hard-headed, empirically-oriented thinker is, in practical terms, setting up a Platonic Heaven by invoking "mere possibility," and thereby ontologically downgrading the realm of the potential. But if we give full ontological status to those possibilities, *we bring them into the same world in which we find sensible particulars.* The language of Form and Field is well-suited to this task.

Anna Marmodoro observes that "the causal efficacy of [Plato's] Forms [is] normativity governing how objects in the world are qualified" (2021, 65), and notes that

> Plato saw in geometry incontestable evidence that the transcendent can govern the physical, on account of the fact that certain features of physical figures may be shown to be true of them on the basis of proofs concerning transcendent figures. This shows that the governance of the physical by the transcendent is possible, and makes Plato's assumption that the transcendent Forms govern the constitution of things in the world plausible. (Ibid., 200)

Thus "Forms are *constitutional* causes, while physical things are efficient causes" (Ibid., 75, emphasis added).

Given the foundational importance of such constitutional, causal relations, they surely ought to be accorded ontological status. If accepting potential material relations as Primordial Forms seems too radical, a bridge too far (as it does for Polanyians such as Jon Fennell)[28] it would be well to recall that such relations were some of the most common examples of Forms in the Platonic corpus as well as to consider that, as we have seen, according to proponents of some currently popular ontological theories (e.g. ontic structural realism) there is ultimately nothing but relations, all the way down. The proposal I am working out is far less radical and quite commonsensical in the sense that I am not interested in dissolving higher-level entities such as medium-sized dry goods into more basic heaps of relations, but rather in recognizing the dry goods' logical dependence on the real existence of such relations, thus giving these their proper due. And again, it is in no way novel to construe relations as Forms, which Plato does himself, as we will soon see.

At any rate, not only do Primordial Forms, by means of their role as that which delineates the spectrum of possible rational causal relations in the material sphere, allow sensible entities to arise, they allow them to be intelligible by means of tacit integration, for, as Lowney points out, "not only are intuitions without concepts blind, there is no intelligible structure to the sense intuition—even in experience—without the senses [*Sinne*] or concepts" (2013–2014, 20). Similarly, Gail Fine notes correctly that "Plato at least sees that we do not first identify particulars, and only later apply general concepts to them; to identify anything, one must

[28] Jon Fennell, comments on a draft, delivered to the author, November, 2019.

already possess general concepts. That is the core of truth in the doctrine of *Anamnesis* that Aristotle's simple account of concept acquisition ignores" (1983, 42 emphasis added).

7 Chapter Conclusion

In this chapter we have explored in some depth Polanyi's metaphysical use of fields, examining objections to his field theory, but concluding that fields are useful analogical metaphysical postulates. Furthermore, I have argued that Polanyi's field theory explicitly and implicitly entails a Platonist position on the existence of Forms insofar as it insists on rational principles as a necessary precursor to the emergence of material entities (thus affirming the existence of uninstantiated Universals, in contemporary analytic terms), something that Polanyi himself did not acknowledge in writing. We have further examined Polanyi's explicitly realist position on the question of Universals and seen how these are best construed within a post-critical framework as comprehensive entities. Given all this, the Platonic nomenclature of Forms was adopted in place of Polanyi's own usage in order to give such Universals their realist and substantive due, acknowledging thus their metaphysical role as the ground of both the natures of participating particulars and the universals we use to track such natures. I have also argued that those Forms which must necessarily be instantiated for physical, sensible reality to develop at all make up a sub-class referred to as Primordial Forms. Finally we laid out the way that Forms populate Polanyi's generative fields, thus nourishing and constraining the development of sensible reality. Along the way questions of concept-formation arose, and it is to these matters that we now turn.

References

Ademollo, Francisco. 2013. "Plato's Conception of the Forms: Some Remarks." *Universals in Ancient Philosophy*. Ed. Riccardo Chiaradonna and Gabrielle Galluzzo. Pisa: Edizioni Della Normale: 41–86.

Aristotle. *Categories* trans. E.M. Edghill. *Internet Classics Archive.* Accessed April 24, 2023. http://classics.mit.edu/Aristotle/categories.1.1.html.

Armstrong, D.M. 2018 [1989]. *Universals: An Opinionated Introduction.* New York, NY: Taylor and Francis.

Bhaskar, Roy. 2008 [1977]. *A Realist Theory of Science.* New York, NY: Routledge.

Bracken, Joseph A. 2004–2005. "Emergent Monism And Final Causality: A Field-Oriented Approach." *Tradition and Discovery* 31/2: 18–26.

Broackes, Justin. 2006. "Substance." *Proceedings of the Aristotelian Society.* Vol. 106: 133–168.

Clayton, Philip. 2002–2003. "Emergence, Supervenience, and Personal Knowledge." *Tradition and Discovery* 29/3: 8–19.

Cohen, S. Marc. 2016. "Aristotle's Metaphysics." *The Stanford Encyclopedia of Philosophy* (Winter 2016 Edition) edited by Edward N. Zalta: https://plato.stanford.edu/entries/aristotle-metaphysics/#SubsUniv

d'Hoine, Pieter. 2011. "Aristotle's Criticism of Non-SubstanceForms and its Interpretation by the Neoplatonic Commentators." *Phronesis* 5/3: 262–307.

Emmett, Dorothy. 1945. *The Nature of Metaphysical Thinking.* London: Macmillan & Co.

Eyring, H. and Polanyi, M. 1931 [2000]. "On Simple Gas Reactions." *Quantum Chemistry: Classic Scientific Papers* trans. & ed. Hinne Hettema. River Edge, NJ: World Scientific Publishing: 423–451.

Franklin, James. 2015. "Uninstantiated Properties and Semi-Platonist Aristotelianism." *Review of Metaphysics* 69: 25–45.

Fennell, Jon M. 2013. "Polanyi, Universals, and the Nominalism Controversy." *The Journal of Speculative Philosophy* 27/4: 365–387.

———. 2017. "Is Polanyi's Emergence Reductive?" *Appraisal* 11/2: 23–36.

Fine, Gail. 1983. "Plato and Aristotle on Form and Substance." *Proceedings of the Cambridge Philological Society, New Series* 29/209: 23–47.

Gerson, Lloyd P. 2005. *Aristotle and Other Platonists.* Ithaca, NY: Cornell University Press.

Grosz, Elizabeth. 2017. *The Incorporeal: Ontology, Ethics, and the Limits of Materialism.* New York, NY: Columbia University Press.

Gulick, Walter B. 1999. "The Meaningful *and* the Real in Polanyian Perspective." *Polanyiana* 8/1–2: http://www.kfki.hu/chemonet/polanyi/; http://www.ch.bme.hu/chemonet/polanyi/

———. 2002–2003. "Response to Clayton: Taxonomy of the Types and Orders of Emergence." *Tradition and Discovery* 29/3: 32–47.

———. 2004–2005. "On Structured Societies and Morphogenetic Fields: A Response to Joseph A. Bracken." *Tradition and Discovery* 31/2: 31–36.

———. 2021. "Michael Polanyi's Understanding of Field Theory." Tradition and Discovery 47/2: 34–45.
Harman, Graham. 2018. *Object-Oriented Ontology: A New Theory of Everything.* London: Pelican Books.
Hart, David Bentley. 2018. "The Mind's Eye: The Conditions of Human Consciousness." Lecture given at the University of South Carolina, 15 November 2018: https://www.youtube.com/watch?v=3bTtd1vCjHs
Kastner, R.E., Kauffman, Stuart, and Epperson, Michael. 2018. "Taking Heisenberg's Potentia Seriously." *International Journal of Quantum Foundations* 4/2: 158–172.
Lowney II, Charles W. 2008. "The Tacit in Frege." *Polanyiana*, 17/1–2: 19–37.
———. 2013–2014. "From Epistemology to Ontology to Epistemontology." *Tradition and Discovery* 26/1: 16–29.
Marmodoro, Anna. 2021. *Forms and Structure in Plato's Metaphysics.* Oxford: Oxford University Press.
Meek, Esther Lightcap. 2017. Contact with Reality: Michael Polanyi's Realism and Why It Matters. Eugene, OR: Cascade Books.
Moodey, Richard W. 2004–2005. "Comments on Joseph A. Bracken's 'Emergent Monism And Final Causality: A Field-Oriented Approach.'" *Tradition and Discovery* 31/2: 27–30.
Mullins, Phil. 2022. "The Journal *Humanitas* as an Incubator of Polanyi's Ideas." *Tradition and Discovery* 48/1: 39–51.
Nagel, Thomas. 2012. *Mind and Cosmos: Why the Materialist Neo-Darwinian Conception of Nature is Almost Certainly False.* New York, NY: Oxford University Press.
Nietzsche, Friedrich. 1979. "On Truth and Lies In A Nonmoral Sense." *Philosophy and Truth.* New Jersey: Humanities Press International, Inc.: 78–91.
Pabst, Adrian. 2012. *Metaphysics: The Creation of Hierarchy.* Grand Rapids, MI: Wm. B. Eerdmans Publishing Co.
Plato. 1997. *Phaedo* trans. G.M.A. Grube, in *Plato: Complete Works.* Edited by John M. Cooper. Indianapolis, IN: Hackett Publishing Company.
Polanyi, Michael. 1962 [1958]. *Personal Knowledge: Towards a Post-Critical Philosophy.* Chicago, IL: University of Chicago Press.
———. 1964 [1946]. *Science, Faith, and Society.* London: Oxford University Press.
———. 1966. *The Tacit Dimension.* Chicago, IL: University of Chicago Press.
———. 1969. *Knowing and Being.* Edited by Marjorie Greene. Chicago, IL: University of Chicago Press.

Polanyi, Michael and Prosch, Harry. 1975. *Meaning*. Chicago, IL: University of Chicago Press.

Salem, Eric. 2017. "Object and Οὐσία: Harman and Aristotle on the Being of Things." *Contemporary Encounters with Ancient Metaphysics*. Ed. Abraham Jacob Greenstine and Ryan J. Johnson. Edinburgh: Edinburgh University Press: 224–242.

Sedley, David. 2013. "Plato and the One Over Many Principle." *Universals in Ancient Philosophy*. Ed. Riccardo Chiaradonna and Gabrielle Galluzzo. Pisa: Edizioni Della Normale: 113–138.

Takaki, Kyle. 2018. "Reality Criss-Crossed." *Tradition and Discovery* 44/3: 15–26.

Aristotle, Plato, and Polanyi on Access to Forms

1 Induction and *Epagōgē*

In general, commentators on Polanyi's engagement with the question of Forms or Universals tend to loosely align him with the family of inductively-oriented views that can be traced back to Aristotle.[1] Phil Rolnick notes that, like Aristotle in *Posterior Analytics* II:19, "Polanyi observes that the problem of forming universals is a problem of empirical induction" (1992–1993, 20). Similarly, Fennell argues that Polanyi's view is that Universals are "certainly derivative of particulars" (2013, 377). Both of these comments only make sense in the context of a set of assumptions that pit Aristotelian Universals against Platonic Forms as rival metaphysical explanations.

As I have already indicated, I follow Gerson (2005) and reject this interpretation as the best way of understanding both Forms and what Aristotle calls universals – our conceptual and cognitive means of tracking similarity-in-difference which are themselves unable to provide a

[1] The exception to this trend, as has already be noted, is Lowney (2008, 2017), who does identify Polanyi as holding a form of Platonism with respect to Universals.

M. E. Turkis II, *The Metaphysics of Michael Polanyi*,
https://doi.org/10.1007/978-3-031-41211-0_6

metaphysical ground for such similarity-in-difference. A further unpacking of what I take to be a proper understanding of Forms and universals will form part of the following chapter. Our immediate concern here is to address the set of objections likely to be raised by thinkers (Polanyian or otherwise) who are, whether explicitly or implicitly, influenced by such an exclusive Aristotelianism. We will do so by considering Aristotle's and Plato's explanations of how we come to acquire concepts. The results, I believe, will show that if we do take these theories as rivals, then Aristotle's theory of Universals is the less compelling, but that if we understand Aristotle as only seeking to address our conceptual tracking mechanism (i.e. universals), then it makes perfect sense within the context of an ampler Platonic metaphysics of Forms, a context that we have already seen harmonizes well with Polanyi's theorizing.[2]

While it is the case that Polanyi's explicit view is that we approach universals inductively, things are far more complicated than this would seem to imply, for, as Polanyi himself argues repeatedly, it is not possible to formulate any explicit or complete set of procedures that necessarily lead to strong inductive inference. If we take seriously Polanyi's epistemology, then, we cannot assume that concept-formation is solely inductive since other forms of reasoning as well as non-inferential intuitions may well figure importantly in its tacit backdrop. As Francisco Gallardo rightly notes, on Polanyi's account induction operates within the knower's more general preconceptions about nature and reality (2003, 304).[3]

Furthermore, the term that Aristotle uses in the passage referenced by Rolnick – *epagōgē,* usually translated as induction – does not map cleanly onto more modern conceptions of inductive reasoning (Hamlyn 1976; Engberg-Pedersen 1979; Mercado 1997), nor is there widespread agreement among scholars of Aristotle on precisely how coherent his

[2] Both Anti-Platonist Aristotelianism and the Platonist Aristotle will therefore make appearances in this chapter. The focus here, however, will be on the influential Anti-Platonist interpretations. Let it be said here once again that better, harmonized interpretations of Aristotle are possible. Such Platonist harmonization will be foregrounded in the following chapter.

[3] "En la inducción influye de manera determinante la preconcepción que el hombre posee de la naturaleza. Detrás de esta consideración se trasluce el papel que Polanyi otorga a la tradición: no se puede separar la tarea de investigación de un científico de la visión de la realidad que ha recibido." (Gallardo 2003, 304)

conception of *epagōgē* is.[4] The kicker, however, is that some interpreters argue that *epagōgē* isn't about concept formation at all, but rather "*about application of a universal to cases*" (Hamlyn 1976, 177 emphasis added), which is clearly out of accord with contemporary understandings of induction. All of this is highly problematic for the commonplace Anti-Platonist, whiggish account of the debate over Universals which holds that

> For Plato, each apple…is but an imperfect example of the absolute (and perfect) Idea of an apple. *But as Aristotle quickly realized, Plato has it exactly backwards*: we arrive at the general idea of 'apple' by mentally abstracting a set of characteristics we think common to all actual apples. *It is we who conjure the 'perfect' idea from the world, not the world copying the concept.* (Pigliucci 2011, emphasis added)

From there it's supposed to be a straight and steady course to the Enlightenment and modern science, with Bacon, Mill, and others furthering this project by tidying up our understanding of induction.

---

[4] Thomas V. Upton sums up such disagreement thus (1981, 172):

> When examining Aristotle's concept of induction, one is faced with two apparently conflicting views. On the one hand…Aristotle's concept of induction is incomplete, ill-conceived, unsystematic and generally unsatisfactory when compared with his theory of deduction. On the other hand…Aristotle's own explicit doctrine…state[s] that induction (*epagōgē*), by being able to grasp the universal (*katholou*) propositions from which demonstrations proceed, is the starting point of all scientific demonstrations effected by categorical syllogisms, most properly in the first figure… In light of the apparent contrast between these two accounts of *epagōgē*, four major possibilities arise which admit that Aristotle states that *epagōgē* is the basis of demonstration but differ on how this is to be interpreted. These possibilities are: (1) either *epagōgē* is in fact incomplete and ill-conceived, but it still can be the basis of demonstration if *nous* is viewed as a faculty that guarantees the truth of premises initially but inadequately gathered by induction;…(2) or epagōgē is incomplete and ill-conceived and therefore by itself it cannot be the basis of demonstration - Aristotle is mistaken in his account of *epagōge*, at least as this account is traditionally interpreted;…(3) or *epagōgē* is not so incomplete and ill-conceived as is commonly held, but it cannot be the basis of demonstration for it yields insufficient results - hence, there is need for a reinterpretation of Aristotle's concept of induction; (4) or *epagōgē* is not so incomplete and ill-conceived as is commonly held and there is the real possibility that it can be the basis of demonstration, for it can lead directly to insight into the universal and necessary causes or natures of individuals investigated.

There are many problems with this account, and I hope in the following pages to point some of them out, especially with respect to what I take to be misapprehensions of Plato and Aristotle[5] and the influence of these misapprehensions on dominant interpretations of Polanyi. The first, and possibly most important shortcoming here is the argument that I have already sketched, following Gerson (2004, 2005) that Plato's Forms are the metaphysical ground both for the properties of the apples *and* for the cognitively-generated universal we use to track apples in the world. Yet another fundamental problem with this account has already been succinctly pointed out by Fine: "to identify anything, one must already possess general concepts" (1983, 42). Without prior (tacit) access to concepts – to Forms, that is, which serve as first principles for the process of perception – there cannot be even the bare recognition of a particular, let alone a process of abstraction which generates a universal from repeated exposure to similar particulars. Why this is so should become clear shortly.

I also wish to point out that the dominant view of concept acquisition amongst scholars of Polanyi, which tends to hew closely to that of Aristotle and to dismiss Plato's account makes it difficult for Polanyi's interpreters to see how close he (Polanyi) is in certain respects to Plato. To that end, let us turn our attention to some important texts on concept acquisition – first, the section of Aristotle's *Posterior Analytics* referenced earlier by Rolnick, and then a portion of the *Phaedo* by Plato – before returning to Polanyi himself.

I will quote Aristotle and Plato at length so as to fairly represent their views, as an antidote to the tendency toward caricature, and also in order to provide direct access to my source texts for those more familiar with Polanyi than with the ancients. While I am most certainly fighting under Plato's banner, I hope that it will become apparent that once we properly grasp certain lacunae in Aristotle's theory of concept-formation and subtleties in Plato's – both of which are commonly overlooked when these two thinkers are not the primary focus of research – a certain harmonization of their positions in light of Polanyi's contributions is not out of the

[5] The misapprehension is of course that the two are in necessary conflict and that Aristotle is best thought of as an Anti-Platonist.

question "since both attempt to find a middle ground between complete possession and complete absence of knowledge" (Adamson 2010, 19).

2 Aristotle

At the beginning of *Posterior Analytics* II:19, Aristotle is inquiring into "the conditions required to produce demonstrative knowledge," and more specifically, how basic premises required for scientific demonstration come to be known – whether they are known innately or learned by other means. He argues thus:

> Now it is strange if we possess them from birth; for it means that we possess apprehensions more accurate than demonstration and fail to notice them. If on the other hand we acquire them and do not previously possess them, how could we apprehend and learn without a basis of pre-existent knowledge? For that is impossible...So it emerges that neither can we possess them from birth, nor can they come to be in us if we are without knowledge of them to the extent of having no such developed state at all. *Therefore we must possess a capacity of some sort,* but not such as to rank higher in accuracy than these developed states. And this at least is *an obvious characteristic of all animals, for they possess a congenital discriminative capacity which is called sense-perception.* But though sense-perception is innate in all animals, in some the sense-impression comes to persist, in others it does not. So animals in which this persistence does not come to be have either no knowledge at all outside the act of perceiving...animals in which it does come into being have perception and can continue to retain the sense-impression in the soul: *and when such persistence is frequently repeated a further distinction at once arises between those which out of the persistence of such sense-impressions develop a power of systematizing them and those which do not. So out of sense-perception comes to be what we call memory, and out of frequently repeated memories of the same thing develops experience*; for a number of memories constitute a single experience. *From experience again – i.e. from the universal now stabilized in its entirety within the soul, the one beside [or over] the many which is a single identity within them all* – originate the skill of the craftsman and the knowledge of the man of science, skill in the sphere of coming to be and science in the sphere of being. (Ibid. emphasis added)

Given that this is a core text for Anti-Platonist Aristotelianism, we will begin by interpreting the passage with the assumption that Aristotle is here offering a metaphysical theory meant to replace Plato's Forms.

Now, since raw sense perception already applies general concepts (such as Limit or Boundary, Equality, Difference, etc.) in the recognition of any differentiable, discriminable entities, then this is to punt on the issue, even if what Aristotle is trying to say is that sense-perception must first be channeled through experience in order to explain our first contact with Universals. After all, he calls sense perception a "congenital discriminative capacity" (ibid.). Yet no discrimination at all is logically possible without recourse to, at the very least, Boundary, Equality, and Difference, all of which are universal concepts which require some metaphysical ground.[6] If, on the other hand, Aristotle is pointing out that perception itself implies prior access to such metaphysical grounds, or Forms, then he is not very clear about this. At this point it is fair to say that if Aristotle is here pursuing a theory of Universals which would supplant Forms as a metaphysical grounding point for the concepts we explicitly acquire through our experience of sensible particulars, then he is falling short. If, however, he is merely addressing the question of our explicit conceptual tracking of consistent properties in sensible particulars, then he may well be on solid ground, as long as some other explanation of the necessary metaphysical grounding were available.[7]

At any rate, Aristotle soon transitions in *Posterior Analytics* to a discussion typically interpreted as explaining, by means of a metaphor, how our sense-perception allows us to acquire concepts:

---

[6] In an email message to the author, November 29, 2019, sociologist Richard Moodey points out that he is "not sure that when an infant discriminates between her mother's face and all the other things in her field of vision that she [is] having recourse to the universal concepts of boundary, equality, and difference." Certainly she is not having *explicit* recourse to these universals, but I insist that she must have *tacit* recourse, for the mother's face is, after all, discriminable. The infant's special predilection for the face of the mother is of course in no way exhaustively explained by the tacit awareness of the Forms here mentioned, however such awareness is necessary though not sufficient for such affordances on the part of the infant, for to have a special, innate capacity to recognize faces and in particular the face of the mother is to have the innate capacity to tacitly carry out logical processes that rely upon such concepts. Thus tacit contact with such Forms is presupposed by such capacities and affordances.

[7] This is, in fact, a persuasive Platonist reading of this passage which fits nicely into a post-critical Platonist synthesis of Plato and Aristotle.

> Let us now restate the account given already, though with insufficient clearness. *When one of a number of logically indiscriminable particulars has made a stand*, the earliest universal is present in the soul: for though the act of sense-perception is of the particular, its content is universal – is man, for example, not the man Callias. A fresh stand is made among these rudimentary universals, and the process does not cease until the indivisible concepts, the true universals, are established: e.g. such and such a species of animal is a step towards the genus animal, which by the same process is a step towards a further generalization. Thus it is clear that we must get to know the primary premises by induction; for the method by which even sense-perception implants the universal is inductive." (*PA* II:19 emphasis added)

Ironically, as Peter Adamson notes when referring to the metaphor of the first universal taking its stand, "the simile is…perfect: but what it perfectly represents is Plato's view, not Aristotle's" (2010, 13). Precisely why this is should become clear soon enough as we look more closely at Plato's position. Again, however, the problem is that the intervention of some Universal concept, some Form, is necessary for the first stand-taker to become logically distinct and discriminable at all. For this reason some see Aristotle's conception of *nous* (intuition) – the only "kind of thought…more accurate than scientific knowledge" which "apprehends the primary premises" (*PA* II: 19) – as the faculty that "guarantees the truth of premises initially but inadequately gathered by induction" (Upton 1981, 172). It must be said, however, that on this view such intuitive powers are only "possible for human beings because their intellectual activity [i.e. *nous*] imitates that of the divine," that is to say, the unmoved mover (Cleary 1993, 13–14). Yet if this is the case, then *nous* is "a generalising capacity" that allows for our conceptual grasp of a universal (Engberg-Pedersen 1979, 308) and it is playing the same explanatory role as Plato's *anamnesis* (or Polanyi's tacit dimension along with his fields, or certain aspects of Peirce's abduction). The point is that there is no unproblematic, purely inductive approach (in the contemporary sense) to the first apprehension of Universals through particulars.

If this is what he is getting at, then Aristotle must see that the circularity of the moment of first recognition puts his emphasis on particulars as primary in a difficult and tenuous position. The answer, however, that somehow, inexplicably (when taken on Aristotle's own terms, from the passage itself, and when interpreted through the lens of contemporary understandings of induction) a lonely particular stands up, simultaneously becoming discriminable and offering the knower her *first* taste of a Universal is entirely too pat.[8] In consequence, Hamlyn concludes that Aristotle never explains how our ability to access and use concepts arises, pointing out that "there appears to be a definite lacuna in the theory here" (1976, 177). Likewise, after an exhaustive analysis, Fine deems Aristotle's account overly simplistic when compared to Plato's (1983, 42). Nonetheless, Aristotle's theory that Universals are inductively acquired (in the modern sense of induction) is often taken to be, at many removes, a devastating refutation of Plato's more nuanced account.

Consider, however, that Aristotle's account makes perfect sense if the universals he is explaining are not construed as metaphysical rivals to Platonic Forms, but rather refer to units in our explicit cognitive tracking mechanism. If the knower has tacit, anamnetic contact with the Forms which ground the nature of the observed particular (say, an apple), then we have some explanation for how it could be possible for the rudimentary universal (i.e. a fallible, cognitively-generated conceptual tracker) to take root upon successive observations of apples, for the Forms metaphysically ground not only the nature of the apples themselves, but also our fallible access to such a nature through the medium of perception. This superior interpretation of the end of *Posterior Analytics* is, roughly speaking, the harmonization of Plato and Aristotle's theories urged by ancient Neoplatonists (Gerson 2005) which we will engage with once more in the following chapter.

3 The Influence of Aristotelianism in Polanyi Studies

Before diving into Plato's view in more depth, it will be useful to look at one example of the influence the Aristotelian account[9] can wield over the interpretation of Polanyi. This is clearly seen in "Polanyi, Universals, and

[8] My analysis here is of course unfair to the historical, Platonist Aristotle. My current critique here is aimed at the common Anti-Platonist Aristotelian account which holds that we simply abstract the immanent Universal (taken to be a metaphysical rival to Plato's Forms) from the particular.

[9] The Anti-Platonist Aristotelian account, that is.

the Nominalism Controversy," in which Jon Fennell argues that Polanyi's position on Universals "overcome[s] the traditional opposition between belief in universals and belief in a world populated solely by particulars...without positing anything like a Platonic form of 'X.'" (2013, 372–373). While Fennell, ostensibly following Polanyi, affirms the reality of the Universal itself, he also stipulates that it "is certainly derivative of particulars" and that the reality of Universals is not *actualized* (2013, 377). Fennell's use of "actual"[10] is in part based on his reading of Polanyi, but is also derivative of Peirce, who,

> in a 1903 essay, directs our attention to "the general idea of dog" (1998c, 222). Through observation of dogs (induction), followed by the exercise of his powers of explanatory hypothesis (what Peirce calls "abduction"), he has "acquired some general ideas of dogs' ways, of the laws of caninity" (1998c, 223). Peirce emphasizes that we come to know dogs and understand their ways not through a fundamentally mysterious grasp of some thing-in-itself behind sensation but, instead, as a product of our mental action; that is, by perceiving certain characteristics and then confirming through experience—through practical effects—that they are actually the case. The universal "dog" is thereby real and meaningful. (Fennell 2006, 43)

We see here an excellent example of Fennell's Polanyio-Peircean take on how a real Universal can be inductively derived from particulars, yet it must be noted that this account is thoroughly Aristotelian, nearly identical to the stand-taking of primitive Universals seen in *Posterior Analytics*, though it has been run through a new set of terms. The particular dogs are, to use Fennell's preferred terminology, both "actual" and "real," while the Universal extrapolated therefrom is "real" but not "actual."

It would seem at this point that perhaps Fennell's position on Universals (which he takes to be Polanyi's) could be quite close to the post-critical Platonist position I am arguing for. It depends on what he means by "actual." Does the Universal of caninity logically pre-exist the instantiation of particular, concrete dogs as a stable, open system as yet uninstantiated, a potential attractor in a gradient of meaning? If so, then Fennell's

[10] Unrelated to Roy Bhaskar's (2008) use of the term.

position *is* quite close to mine, however it would be a Platonist position, and as we have seen above, Fennell explicitly rejects Platonism.[11]

Thus "actual" in Fennell's usage cannot mean simply "instantiated" in a Platonic sense. Indeed, it seems that not only in this example, but also in Fennell's other relevant writings "actual" is a placeholder for "empirical" (in the current scientific sense) and "particular." The postulation of non-empirical Universals is thus supposed to be a novel evasion (by Polanyi and Peirce) of the nominalist menace. But of course Forms are not empirical objects for Plato either, though they are certainly real. The nominalist, then, is unlikely to be appeased by Polanyi if she was unmoved by Plato (or Aristotle). In addition, Fennell's terminological combination "real and actual" maps quite cleanly, so far as I can tell, onto Aristotle's conception of primary substance, while "real but not actual" stands in for secondary substance (*Categories*, Part V).

Returning to the example of the dogs while replacing "actual" with "empirical," to say that particular dogs are empirical and real while the Universal is real but not empirical is hardly a new or surprising argument for almost any realist with respect to Universals, antique to present. Furthermore, when Fennell echoes Peirce in celebrating our avoidance of any "mysterious grasp of some thing-in-itself," (Ibid.) he fails to deal with the fact that, again, to recognize dogs as comprehensive entities at all and thus as members of a class hitherto unknown to us we must be accessing and using concepts (Limit, Boundary, etc.) that are real entities unto themselves, without which our integrative, abductive powers collapse.

I understand the hesitation in postulating a necessarily existing Form of the Canine as Primordial in order to account for the contingent existence of particular dogs. My post-critical Platonism handles this intuitive hesitation by distinguishing Primordial Forms (again, those Forms whose instantiation is a requirement for material reality to exist at all) from non-primordial Forms, which may or may not be instantiated in any possible world with material components. The Form of the Canine, like the Form

[11] Thus, for clarity's sake, Fennell's account is similar to the Aristotelianism that holds to the interpretation of *Posterior Analytics* as positing a metaphysical theory of Universals that should replace Plato's theory of Forms and *not* a theory of universals that would augment the theory of Forms.

of any other plant or animal, or artifact like a bicycle, is most certainly not primordial in this sense.

Perhaps it is the case that knowledge of such non-primordial Forms can be accessed inductively, with no tacit, anamnetic, prior contact *with the non-primordial, contingently instantiated Form in question* (e.g. Dog, Birch, Bicycle, etc.). I am not sure how we would ever find out. It is crucial, however, that we note here the necessity of tacit, anamnetic prior contact *with Primordial Forms* (e.g. Equality itself, Division itself, etc.) in order to account for any deductive, inductive, or abductive proceedings within the sensible world whatsoever, including those leading to our recognition of diamonds as diamonds, dogs as dogs, birches as birches, bicycles as bicycles. It is precisely for this reason that non-primordial Forms cannot serve appropriately as paradigm cases for our understanding of concept formation.

4 Plato

In light of this, let us consider now, in contrast to the Aristotelian approach to concept formation presented at the end of *Posterior Analytics* (along with its Polanyian and Peircean variants), Plato's presentation at *Phaedo* 74b-75a of the same juncture, the moment at which explicit inductive epistemological contact is made with a universal concept, arriving by means of a particular in an already-discriminable state:

> Whence have we acquired the knowledge of [likeness or Equality]? Is it not from the things we mentioned just now, from seeing sticks or stones or some other things that are equal we come to think of that other which is different from them? Or doesn't it seem to you to be different? Look at it also this way: do not equal stones and sticks sometimes, while remaining the same, appear to one to be equal and to another to be unequal?
>
> Certainly they do.
>
> But what of the equals themselves? Have they ever appeared unequal [c] to you, or Equality to be Inequality?
>
> Never, Socrates.
>
> These equal things and the Equal itself are therefore not the same?

I do not think they are the same at all, Socrates.

But it is definitely from the equal things, though they are different from that Equal, that you have derived and grasped the knowledge of equality?

Very true, Socrates.

Whether it be like them or unlike them?

Certainly.

It makes no difference. As long as the sight of one thing makes you think of another, whether it be similar or dissimilar, this must of necessity be recollection?

[d] Quite so.

Well then, he said, do we experience something like this in the case of equal sticks and the other equal objects we just mentioned? Do they seem to us to be equal in the same sense as what is Equal itself? Is there some deficiency in their being such as the Equal, or is there not?

A considerable deficiency, he said.

Whenever someone, on seeing something, realizes that that which he now sees wants to be like some other reality but falls short and cannot be [e] like that other since it is inferior, do we agree that the one who thinks this must have prior knowledge of that to which he says it is like, but deficiently so?

Necessarily.

Well, do we also experience this about the equal objects and the Equal itself, or do we not?

Very definitely.

We must then possess knowledge of the Equal before that time when we first saw the equal objects and realized that all these objects strive to [75] be like the Equal but are deficient in this.

That is so.

Then surely we also agree that this conception of ours derives from seeing or touching or some other sense perception, and cannot come into our mind in any other way, for all these senses, I say, are the same.

They are the same, Socrates, at any rate in respect to that which our argument wishes to make plain. (emphasis added)

First, we ought to note that Plato takes here as a paradigm case of concept formation not the non-primordial sticks or stones themselves, nor yet a paradigmatic Form of Stick or Stone, but rather the relation of Equality, which is, in my view, a Primordial Form. David Sedley, commenting on Plato's choice of example, notes that

> Another thing we know about equality is presumably that it is a transitive property: if A is equal to B, and B is equal to C, A is equal to C. Arguably, even if you have never heard of the word 'transitive', and have never formulated this principle to yourself, you already know it to be true. Indeed, you could hardly count as a rational being if you did not. But – and this is what makes the choice of equality such a felicitous one for Socrates' current argument-you cannot possibly have got that knowledge about equality from sense perception. The senses provide many apparent counter-examples to the transitivity of equality. In a group of three bottles, bottle A looks equal to bottle B and bottle B looks equal to bottle C, but bottles A and C do not look equal. Yet no amount of such sensory counter-evidence could ever weaken your belief in the transitivity of equality, because you already know its truth independently of sense experience…what we know about it intellectually is such as to override the apparent counter-evidence of the senses. (Sedley 2007, 71–72)

Thus dogs, bicycles, mountains, or anything else that is similar insofar as it is a contingent, composite, material entity cannot serve as a proper example of fundamental access to Universals since such kinds cannot play the role described by Sedley above. Such kinds could, theoretically at least, be accessed inductively *as long as* the knower accessing them has prior, tacit knowledge of the more basic, fundamental concepts (Primordial Forms) necessary for the decoding and recognition of the composite structure of the new category. In the search for metaphysical grounds, Plato's choice of a paradigmatic case is felicitous indeed.

In my estimation, then, this passage from *Phaedo* shows far more nuance than Aristotle's account of concept-formation insofar as Plato does not skate over the mysterious nature of our recognition of equality in the sticks. It is thus a clear example of what Charles Lowney is getting at when he discusses the way that "various particulars can be gathered up into a coherent ideal or a comprehensive entity that is a discovery and not merely a construct" (2017, 162). Plato clearly affirms here that our *focal epistemological approach* to the Form is primarily inductive in that the Form is recognized *only through the sensible participating particulars.* Thus Adamson affirms that Plato's view as expressed in the *Phaedo* is "compatible with Aristotle's claim that it is the *dunamis* of sense-perception that gives us knowledge. Indeed Plato emphasizes that there is no other way…to reach knowledge of the Equal than by sense-perception" (Adamson 2010, 8).

Yet paradoxically (and this is precisely where Plato's view is in my view clearly superior in its forthright confrontation with the paradox of our aporetic finitude), for this access to be possible at all – that is for the particular to be discriminable as a particular participating in the Form – we must logically have had *tacit* access to the Form in some way *prior* to our contact with the particulars.[12] Plato's proposed solution is the soul's prior knowledge of the Form from past lives, to which one might respond that he "is disposing of an enigma by postulating a miracle" (*KB* 167).

But of course this is the skeptical response anticipated by Polanyi himself to *his* own philosophical proposals on the problem of Universals. And it is this same difficulty that leads Polanyi, like Plato before him, to acknowledge the aporia presented by our experience of concept formation, experiences of which we become aware *in medias res,* when he (Polanyi) avers that

> those who insist on finding a formal procedure of induction would reject the acknowledgment of such powers of discovery as mystery mongering. Yet these powers are not more mysterious than our powers of perception but of course not any less mysterious, either (Ibid.).

I would add that these powers are no more nor less mysterious or exotic than Platonic Forms.

5 Polanyi and Plato

Note, however, the way in which Polanyi's epistemology wonderfully compliments this passage from the *Phaedo* if we substitute our tacit integrative capability for Plato's *anamnesis* in order to "resolve the apparent contradiction involved in taking an aggregate of objects which differ in every particular to be…identical in some…way" (Ibid.). The tacit dimension is a plausible and less mythological confrontation with the

[12] As Polanyi also notes in his analysis of the *Meno.*

impossibility of exhaustively outlining our access to the metaphysical resources that justify the rational relation of *being equal to.* Yet we know we must have access to such resources, because we clearly recognize such intelligible properties.

This is a fruitful interplay, indeed. Yet Polanyi's own references to Plato unfortunately do not indicate that he fully appreciated the closeness in this respect of his own project to Plato's. Nor do they demonstrate any especially deep understanding of the complexity and richness of the theory of Forms as debated and unveiled across Plato's corpus. Take, for example, the following passage:

> Plato was the first to be troubled by the fact that *in applying our conception of a class of things, we keep identifying objects that are different from each other in every particular.* If every man is clearly distinguishable from another and we yet recognize each of them as a man, what kind of man is this, as which all these men are recognized? Plato concluded that the general idea of man refers to a *perfect man* who has no particular properties, and of whom individual men are imperfect copies, corrupted by having such properties. (*KB* 165–166 emphasis original)

Though brief and not intended by Polanyi as a comprehensive statement on the theory of Forms, this formulation exhibits problems which are indicative of common, quasi-Aristotelian misunderstandings. The phrase "a perfect man" may seem innocent enough, but it posits the Form in non-analogical, spatio-temporal terms *sans* caveat. This belies a certain degree of confusion, since Forms, properly understood, are "incorporeal, changeless, intelligible realities" which serve as paradigms or attractors, appearances of which are immanent in sensible particulars "as their determining properties" (Perl 1999, 361, 342).

Perhaps we can say that Polanyi's representation here only unintentionally distracts us from the Form's incorporeality, which he well understood. Even so, his reference to the participating particulars as "imperfect copies" is also potentially problematic, for to have a copy (even an imperfect copy) of something is to have another instance of that thing. But of course (recurring again to Polanyi's formulation) it is difficult to see how particular men – material, concrete sensibles – can be copies of the immaterial paradigmatic Form (they are, for one thing, material, at least in part). A better formulation (about which more anon) would be to use the language of appearance and reality (as Plato often does) and thus construe the particular men as sensible appearances of the Form of man. What Polanyi is doing – perhaps unintentionally – as he attempts to understand and represent Plato's view, is implicitly attributing to Plato a reified Two Worlds theory, whereby the Platonic conception of a Form is recast non-metaphorically in spatio-temporal terms, thus creating confusion regarding the way sensible particulars relate to the Forms. Thinking of the Forms this way nudges the unsuspecting interpreter (like the young Socrates of *Parmenides*) towards understanding Forms as "(checkmate)…object[s] in another world" (Robjant 2012, 59). Thus arises the dubious notion of a Platonic Heaven (again, conceived of spatio-temporally as a *place*) where the Forms supposedly reside.

As this misunderstanding of Plato is typically carried forward by interpreters in the modern tradition, the waters are muddied further by Kant's distinction between the noumenal and phenomenal, which metastasizes into a division between the immanent and the transcendent, with the transcendent eventually ruled rationally out-of-bounds due to its noumenality.[13] At this point, the problematic tendency can evolve from a rejection of a reified Two Worlds theory of Forms (taken to be the paradigm case of transcendence) to a knee-jerk metaphysical reaction against the possibility of any transcendent entities whatever, since these are not explicitly articulable phenomena in scientific terms (or, in the context of

[13] See "The End of Metaphysics" by H.O. Mounce:

> [Kant] argued that it is only the world as known, or phenomenal, which is immanent in, or internally related to, the categories of mind. The world itself, or in itself, is transcendent or noumenal. In short, it is radically distinct from the categories of the mind. As such it cannot be known at all. Here, again, is a genuine dichotomy. The transcendent and immanent are so defined as to exclude one another.

Now that dichotomy is entirely alien to the [classical] metaphysical tradition. In that tradition, the human mind is internally related to the world in the sense that it necessarily depends upon it. But the world is not necessarily dependent on the human mind. Rather it transcends it. (2005, 522)

Takaki worries about "conflat[ing] Polanyi's realist assumption with metaphysical realism…[with] its imported baggage of independence" (2018, 21). On Mounce's view, however, it is the phenomenal/noumenal divide that imports such baggage, not metaphysical realism.

some critical theory, due to the assumption that affirmation of any sort of transcendence leads to oppression).[14]

This last stage in the metastasis is certainly not operative in Polanyi's own project, which in fact aims to show in epistemological terms that science is impossible without such inarticulable transcendent entities. Rolnick thus rightly sees clear connections between Plato and Polanyi noting that "Polanyi does not differ from Plato so much in what they believed, but in how the beautiful, true, or good might be approached and transmitted" (1992–1993, 16). This is spot on – however it may be that, given the pervasive misunderstandings of the Forms sketched above, Polanyi is much closer to Plato than either he or Rolnick have recognized.

Thus I think another interpretive path lies open to us here. Polanyi certainly emphasizes our epistemological approach to the Universal through the tacit integration of the subsidiary members of its class, as we have seen. But he is also, I think, intuitively aware of the need to ground certain concepts logically prior to the advent of even the meanest matter in order to provide for its mere intelligibility. After all, intelligibility just is the only possible horizon of existential possibility if one accepts, as I do, that Polanyi is correct in thinking that our epistemological situation is a reliable model for ontology and that therefore meaning is being.

Again, it is, I believe, some form of this intuition that leads Polanyi back time and again to his primordial fields of potentiality, which serve as the anchor for intelligibility and hence meaning. The positing of these fields is therefore, in my view, a move that is actually quite close to the positing of a Platonic Form of Forms (i.e. the Good), which draws the unfolding of actualization onward. Furthermore, this element of the Polanyian imaginary pulls together the transcendent and the immanent in a very Platonic way.

I would argue, then, that Polanyi's fields are best understood as the network of Forms along with the shaping power they exert over the unfolding of actualized reality (in other words, their functioning as

[14] For more on the influence of Anti-Platonist Aristotelianism on twentieth century interpretations of Plato's theory of Forms see Norio Fujisawa (1974, especially 38–40 and on). The passage from Pigliucci (2011, cited earlier) is an example of such influence continuing to color contemporary engagement with Plato *even when* the central topic of the piece is Pigliucci's own conversion to belief in mathematical Platonism. Meanwhile, Gerson (2005) details the complicated interconnectedness of Platonic and Aristotelian thought in the classical world.

constitutive causes). In bringing Polanyi's fields together with Plato's Forms, then, I hope to bring some additional modicum of clarity to this fundamental element of Polanyi's metaphysics. Put simply, Polanyi's account requires Plato's if it is to be as complete and persuasive as possible, and this synthesis shouldn't be terribly difficult to work out, as Polanyi is much closer to Plato than he and most of his interpreters have realized.

6 Chapter Conclusion

We have seen that what is generally taken to be Aristotle's account of concept formation – that we arrive at the first recognition of a Universal inductively after repeated exposure to participating particulars – is unconvincing since it offers no explanation of how a knower might recognize any particular as discriminable from the anything else in the first place. It thus offers no way for knowing to get off the ground, since no knowledge is possible without prior access to concepts. We have also seen, in the cases of Rolnick's and Fennell's respective analyses, how Polanyi's position on Universals is typically interpreted as being of a piece with Aristotle's. I have problematized this interpretive trend firstly by contrasting it with Plato's more nuanced account of access to Forms in the *Phaedo* (arguing thus that the purely inductive theory cannot work either for Aristotle or for Polanyi) and secondly by pointing out that Polanyi's insistence on the ubiquity of the tacit coefficient of knowing allows space for non-inductive elements to work in the process of concept formation (e.g. in the move from tacit awareness of the Form in itself to explicit awareness of the Form in and through the participative particular). The tacit dimension of knowledge thus functions as an updated theory of *anamnesis* for the post-critical Platonist.

References

Adamson, Peter. 2010. "*Posterior Analytics* 2.19: A Dialogue with Plato?" *Bulletin of the Institute of Classical Studies* Supplement 107: *Aristotle and the Stoics "Reading" Plato:* 1–19.

Aristotle. *Categories* trans. E.M. Edghill. *Internet Classics Archive.* Accessed April 24, 2023. http://classics.mit.edu/Aristotle/categories.1.1.html.

_____. *Posterior Analytics* trans. G.R.G. Mure. *The Internet Classics Archive*. Accessed September 24, 2019. http://classics.mit.edu/Aristotle/posterior.html.

Bhaskar, Roy. 2008 [1977]. *A Realist Theory of Science*. New York, NY: Routledge.

Cleary, John J. 1993. "*Nous* as the Ground for Aristotle's Metaphysics?" *The Society for Ancient Greek Philosophy Newsletter* 164: Accessed September 24, 2019. https://orb.binghamton.edu/sagp/164.

Engberg-Pedersen, T. 1979. "More on Aristotelian Epagoge." *Phronesis* 24/3: 301–319.

Fennell, Jon M. 2006. "Jaffa, Nominalism, and the Fate of the Common Noun," *Perspectives on Political Science* 35/1: 37–46.

———. 2013. "Polanyi, Universals, and the Nominalism Controversy." *The Journal of Speculative Philosophy* 27/4: 365–387.

Fine, Gail. 1983. "Plato and Aristotle on Form and Substance." *Proceedings of the Cambridge Philological Society, New Series* 29/209: 23–47.

Fujisawa, Norio. 1974. "Εχειν, Μετέχειν, and Idioms of 'Paradeigmatism' in Plato's Theory of Forms." *Phronesis* 19/1: 30–58.

Gallardo de la Torre, Francisco. 2003. "La Epistemología de Michael Polanyi: Una perspectiva realista de la ciencia." PhD diss., Pontificia Universitas Sanctae Crucis.

Gerson, Lloyd P. 2004. "What is Platonism?" *Journal of the History of Philosophy*. 43: 253–276.

———. 2005. *Aristotle and Other Platonists*. Ithaca, NY: Cornell University Press.

Hamlyn. D.W. 1976. "Aristotelian Epagōgē." *Phronesis* 21/2: 167–184.

Lowney II, Charles W. 2008. "The Tacit in Frege." *Polanyiana*, 17/1–2: 19–37.

———. 2017. "Overcoming the Scientistic Imaginary." *Charles Taylor, Michael Polanyi, and the Critique of Modernity* ed. Charles W. Lowney II. Palgrave Macmillan. 143–168.

Mercado, Juan Andrés. 1997. "La argumentacion retórica como base interpretativa de la inducción aristotélica." *Tópicos* 12: 43–54.

Mounce, H.O. 2005. "The End of Metaphysics." *New Blackfriars* 86/1005 (September): 518–527.

Perl, Eric D. 1999. "The Presence of the Paradigm: Immanence and Transcendence in Plato's Theory of Forms." *The Review of Metaphysics* 53/2: 339–362.

Pigliucci, Massimo. 2011. "Mathematical Platonism." *Philosophy Now* 84: Accessed April 24, 2023. https://philosophynow.org/issues/84/Mathematical_Platonism.

Plato. 1997. *Phaedo* trans. G.M.A. Grube, in *Plato: Complete Works.* Edited by John M. Cooper. Indianapolis, IN: Hackett Publishing Company.
Polanyi, Michael. 1962 [1958]. *Personal Knowledge: Towards a Post-Critical Philosophy*. Chicago, IL: University of Chicago Press.
———. 1969. *Knowing and Being.* Edited by Marjorie Greene. Chicago, IL: University of Chicago Press.
Polanyi, Michael and Prosch, Harry. 1975. *Meaning*. Chicago, IL: University of Chicago Press.
Robjant, David. 2012. "The Earthy Realism of Plato's Metaphysics, or: What Shall We Do with Iris Murdoch?" *Philosophical Investigations* 35/1: 43–67.
Rolnick, Phil. 1992–1993. "Polanyi's Progress: Transcendence, Universality, and Teleology." *Tradition and Discovery* 19/2: 13–34.
Sedley, David. 2007. "Equal Sticks and Stones." *Maieusis. Essays in Ancient Philosophy in Honour of Myles Burnyeat.* Ed. D. Scott. Oxford: Oxford University Press: 68–86.
Takaki, Kyle. 2018. "Reality Criss-Crossed." *Tradition and Discovery* 44/3: 15–26.
Upton, Thomas V. 1981. "A Note on Aristotelian *epagōgē*." *Phronesis* 26/2: 172–176.

Post-Critical Platonism

1 Platonism and the Post-Critical Approach

Given that this project is primarily an intervention in Polanyi studies and that many Polanyi scholars evince some reluctance about Platonism, some clarification of the sort of Platonism I am advocating is in order. This is especially important given the frequency with which the term "'Platonism' as used by modern philosophers…stand[s for]…a theory, which…no one may have actually held" (Gerson 2005, 253). He argues that

> What is most distinctive about Platonism is that it is resolutely and irreducibly "top-down" rather than "bottom-up." A top-down approach to philosophical problems rejects and a bottom-up approach accepts the claim that the most important and puzzling phenomena we encounter in this world can be explained by seeking the simplest elements out of which these are composed.
>
> The top-down approach appeals to irreducible, intelligible principles to account for these phenomena…[and] holds that answers to questions about these phenomena are never going to be satisfactorily given in terms

M. E. Turkis II, *The Metaphysics of Michael Polanyi*,
https://doi.org/10.1007/978-3-031-41211-0_7

> of, say, elementary physical particles from which things "evolve" or upon which the phenomena "supervene." According to this position, "Platonism" is ur [i.e. original, earliest, primal] "top-downism" and its authentic opposite is ur "bottom-upism." Varieties of "bottom-upism" are practically coextensive with varieties of materialism. By "materialism" I mean, basically, the position that holds that the only things that exist in the world are bodies and their attributes, however the latter be construed. All materialists, that is, all anti-Platonists, share the view that, even if attributes are taken to be immaterial in the anodyne sense that they are real and that they are not themselves bodies, they are dependent upon bodies for their existence and explicable entirely in materialistic terms. Thus, for the materialist there are no immaterial or incorporeal entities. (Ibid., 258–259)

I think that Polanyi is uncontroversially a Platonist by these lights. We have already seen that he unflinchingly acknowledges immaterial entities and that, indeed, the goal of his epistemological work was to restore our ability to make such affirmations openly and in good faith. He also explicitly acknowledges the principle of top-downism in his theorizing of fields and gradients, as well as in his work on dual control, as we see when he argues that

> Since our knowledge of... a particular comprehensive – i.e. 'molar' – achievement...is not specifiable in 'molecular' terms, the function itself is not reducible to molecular particulars; *it must be acknowledged therefore as a higher form of being, not determined by these particulars.* We can reach this conclusion directly by recalling that the understanding of a whole appreciates the coherence of its subject matter and thus acknowledges the existence of a value that is absent from the constituent particulars. (*PK* 327 emphasis added)

Gerson notes that for many Middle and Neoplatonists, Aristotle himself was considered to be a type of Platonist, developing the inheritance of the Academy in a particular direction. In this context Platonists generally "saw no impediment to drinking from the font of Aristotelian wisdom in order to understand Platonism better" (Gerson 2005, 259) and some even held that "Aristotle was authoritative for the sensible world and Plato for the intelligible world" (Ibid., 257). This is in stark contrast to many contemporary discussions in which Platonism and Aristotelianism are taken to be mutually exclusive, incompatible positions, as we have already seen in the context of debates over the possibility of uninstantiated Universals.

To read Plato and Aristotle as compatible with one another is not to say that there are no significant points of disagreement, as we saw when looking at their respective views on concept acquisition. Nevertheless, in my view the common understanding of Plato and Aristotle as starkly incompatible and opposed to one another in a fundamental sense has overly swayed the sensibilities of the philosophical world at large, and that includes many interpreters of Polanyi's work. Of particular importance for the current discussion is the way it has shaped debates over the separation of the Forms and Two Worlds Theories.

The thrust of my concluding chapter, then, will continue to argue for a synthesis of the contributions of Plato and Polanyi in the context of the post-critical Platonist metaphysics I have argued for in earlier chapters. It will also gesture toward the fact that Aristotle has much to contribute to this synthesis. Since examining the relationship of Polanyi's thought to contemporary realisms in Chap. 3, two Aristotles (or better, Aristotelianisms) have appeared in the text. Up to this point, we have primarily dealt with the Aristotle of Anti-Platonist Aristotelianism who appears in many contemporary discussions and who is taken to be (as a mature thinker, at least) fundamentally at odds with Plato's theory of the Forms. When Aristotle appears in this final chapter, however, it will be as he was read by many ancient Platonists, which is to say as a particular type of Platonist in his own top-downism.

In the interest of outlining the ways in which misunderstanding of Plato's thought has prevented a fertile cross-pollination from taking place within the context of Polanyi studies, I now aim to offer some further arguments for the compatibility of Platonism with the post-critical approach to philosophy. I will then briefly offer an alternative reading of Plato's theory of Forms, rooted in a well-established interpretive tradition, tapping into what I see as Plato and Polanyi's shared intuitions about transcendence and immanence, and harmonizing quite closely with Polanyi's insights into the tacit dimension of knowing. This will, I trust, helpfully expand upon my proposal of Forms as constituent parts of Polanyi's fields and gradients, operative within the tripartite metaphysics I have developed over the course of this project. Part of this discussion will involve a brief consideration of whether Forms and universals are postulated to do the same metaphysical job as well as a possible rapprochement between Plato's insistence that uninstantiated Forms exist and are substantive and Aristotle's stipulation that no uninstantiated universal exists.

Before beginning this task, I should note that there is an incredibly complex secondary literature treating countless debates over every possible aspect of Plato's theory of Forms. My intention here is not to argue that all scores are settled on such fronts, but rather to mark out what may seem to some Polanyi scholars a more winsome approach to the Forms, rigorous yet nuanced, that has great explanatory power when applied to Polanyi's metaphysics. In so doing I aim to take both Plato and Polanyi seriously in the sense recommended by Stephen R.L. Clark when he argues that taking a philosophical tradition seriously requires that we move beyond textual criticism of the canonical works themselves and explore and develop the implications of such a tradition of thought for those of us here and now, given our recognition that it "is still of value…because it can still inspire its disciples in ways that Plato [or, I would add, Polanyi] himself did not write down, nor even (perhaps) imagine. (2015, 250)

It is important to recognize that despite the fact that Plato is often vaguely held to be fustily dogmatic, he is in reality one of the least dogmatic thinkers on record. His near-constant use of the dialogue form, along with the shift away from Socrates as the primary speaker over time makes it notoriously difficult to pin down precisely which theory of Forms was being affirmed by Plato and when. Furthermore, Plato on the page is less attached to certainty than much received wisdom would indicate. This is an important point for post-critical thinkers who take as a starting point the assumption that non-circular certainty is unavailable to us, but that this recognition does not entail relativism or nihilism.

When I read the *Theaetetus (Plato 1997d)*, for example, I find the parallels to Polanyi's epistemology striking and deep. Note first that the dialogue's purpose – to find an explicit and defensible definition of knowledge – is ultimately fruitless, with the discussants considering and then rejecting the possible definitions of knowledge as perception, as true opinion, and as true opinion plus explicit rational explanation. Furthermore, in the course of the discussion the focus shifts from the objects of knowledge as possible definitional sources to the knowing subject's noetic activity, a move which matches Polanyi's own emphasis on our experience as knowers as the most promising foundation for a proper epistemology. The aporia here stands in as a placeholder for Polanyi's later insights into the impossibility of reducing knowledge to an explicit definition or process. The tacit dimension does not overcome the aporia, but rather acknowledges it as constitutive of a post-critical epistemology and of our actual, lived experience. Not only this, but the investigation carried out by Socrates of the possibility of rationally defining compounds in terms of the simples of which they are composed reads like an ancient anticipation of the theory of tacit knowing.

Returning to the Forms, one significant narrative in the scholarship holds that in an early phase Plato emphasized the Forms' immanence, later shifting to a view of the Forms as separate and austerely transcendent, especially in the *Timaeus (Plato 1997e)*.[1] Related to this interpretive stance is the view that the *Parmenides*, in which the dialogue's namesake subjects a teenage Socrates to withering scrutiny over his hastily formulated theory of Forms is a repudiation by Plato of some of his earlier views. This tradition of scholarship attempts to scrupulously catalog and compare different views of the Forms as they appear across the corpus and to give an evolving historical and authorial through-line to this "development." Yet there are other viable approaches and in light of these it is also clear that an embrace of Forms as explanatorily necessary need not commit one either to the doctrine that the Forms are separate from their particulars in the sense implied by a reified Two Worlds Theory,[2] as I have already alluded to. Sandra Peterson points out that the dialogues include references to many different ways of understanding the Forms and that since many of these interpretations are put forward by characters other than Socrates, one is under no obligation to offer a harmonization or to assume that Plato himself held conflicting views. In light of this,

[1] See Norio Fujisawa (1974) for further discussion of this view, which he accepts in modified form.

[2] See, for example Eric Perl (1999) as well as Gail Fine's essays "Separation" and "Immanence" (2003).

"the *Parmenides*' serious objections to one theory of forms give…no motive to reconcile the objections with the alleged doctrine of Plato" (Peterson 2008, 408).

Yet Peterson also notes that we still have the worthwhile question not only of what Plato believed about the Forms, but also what is worth believing about them ourselves, perhaps guided, to some extent at least, by what we think Plato believed. Her recommended method is the one I advocate here: We have to carefully read the dialogues, though it "is exceedingly hard to tell exactly what [Plato] believes, either from what he says in persuasive mode or from what he asks in questioning mode." From there, under the assumption that the "character Socrates and the author Plato believed about the forms what best survives examination," (Ibid.) we must work, via careful reading and thinking, toward such an examination, imperfect though it may be. As Clark notes, our consideration does not bind us to Plato's texts, and this produces an inescapable tension that "can be negated only by abandoning the whole Platonizing enterprise (2015, 250).

The results of my own ongoing examination of Plato's Forms, blended with Polanyi's language of fields and gradients, include a conviction that, following Mounce and Lowney, we must move away from the hard immanent-transcendent dichotomy (implicitly functioning as phenomenal-noumenal) bequeathed to us by modernity. If we are able to navigate this hazard, then radical, unappealing Two World interpretations of Plato melt away, and we are freed up to see, with Plotinus (1968-88), that on Plato's view "sense perceptions are dim intellections, and intellections…are clear sense-perceptions" (*Enneads* VI.7.7). That is to say that Plato's infamous disdain for the physical, the body, the realm of becoming is all too often overstated and that what is really at stake for Plato is "the view that rational intelligibility is at the root of the natural order" (Nagel 2012, 17) – precisely what is necessary to satisfy the metaphysical aims of Polanyi's fields, namely that there be "some point to the way things are put together" (*M* 161) in order that the fact/value distinction might fall, and also in order that epistemology might plausibly render worthwhile ontological guidance. There are, therefore, "not two worlds, one phenomenal, the other noumenal. There is only one world which, though it transcends or goes beyond human experience, is not radically separate from it. Indeed it is only through this world that there is sense in human experience" (Mounce 2005, 525).

The post-critical Platonist project I am developing thus holds "that the everyday world is very much there, and needs to be brought under rational scrutiny" (McCabe 1994, 26). The Forms I posit as aspects of Polanyi's fields are, then, real properties that "carve up each kind...along its natural joints," as Plato (1997b) puts it in the *Phaedrus* 265e. They are simultaneously transcendent and immanent, as Eric Perl ably explains:

> What Plato presents...is not...a world of sensible instances on the one hand and a world of transcendent forms on the other, but rather one world, that of intelligible form, and the appearances of that world which constitute sensibles. It is this understanding of the properties of instances as appearances of the forms that enables us to reconcile immanence and transcendence. The form is in the instance in that it appears and may be cognized here, in this particular association, as the property of this instance; and it is transcendent, that is, is not itself conditioned by the particularities of this or any other appearance and can be known apart from them. Since the same form may be apprehended as it is, as one, itself by itself, or as it appears, as many, as the property of each of the instances, it is both transcendent and immanent. The transcendence of the forms, then, so vividly presented in the middle dialogues, is not the separation of one world, one set of objects, from another, but is rather the priority and independence of intelligible reality in relation to the sensible appearance of that reality, and is thus a transcendence which does not contradict but rather both implies and is implied by immanence. (1999, 351)

Translating this into the terms of post-critical Platonism, we might say that the Form as one, itself by itself, is the Form as transcendent joint meaning – the Form, that is to say, as comprehensive entity under the focal gaze. The properties of participating sensible instances – the appearances – are Form as subsidiary particular – that is to say the Form as indwelt and harnessed in the comprehensive structure of the participating particular to provide its intelligibility.

This post-critical reformulation of Perl's understanding sets us up to see clearly that no reified Two Worlds Theory is necessary for a robust interpretation of a theory of Platonic Forms. The immateriality and especially the transcendence of Forms construed as comprehensive entities does not derive from their "location" in some alternate reality or state of affairs, but rather from their ontological status as joint meanings which are irreducible to their subsidiary instantiations, even as they are structurally connected ontologically and mereologically.[3] The terminology of Platonic Heaven, a non-spatiotemporal "place" is thus terribly misleading. The "world" or "realm" of the Forms *just is* coextensive with the real; where reality is sensible, it is necessarily shot through with transcendent Form as the metaphysical ground of its immanent and sensible properties. As Iain Hamilton Grant puts it, this is a "one world Platonic metaphysics" (2006, 41).

Understanding the transcendent aspect of the Forms as joint meanings in terms of reality and its immanent manifestation in sensibles as appearance (rather than two separate realities or worlds) is helpful on at least two fronts. First, it is firmly rooted in our everyday experience (as Iris Murdoch puts it "the ordinary way is the way") (1993, 508). We constantly interact with multiple appearances or manifestations of a variety of realities in a fundamentally non-mysterious fashion: Abridged, unabridged, audio, and film versions of *Moby Dick*; chronologically-arranged photo albums that trace a family history, etc.

In addition to this, the reality/appearance relationship maps cleanly onto Polanyi's own Eleatic application of the IFM test of reality, which renders his proper conclusion that immaterial entities such as the mind (and/or personality) of a person are more real metaphysically (i.e. exercise greater Eleatic force) than her biology in that its future (sensible)

---

[3] Here I build on Lowney (2017) who was the first (to my knowledge) to explicitly address these topics in the context of Polanyi Studies. He says,

> The right distinction to make is not between two unbridgeable substances (Descartes) or a phenomenal and noumenal realm (Kant) or an immanent world and a transcendent beyond (in Taylor), or a material world and nonsense (in the scientistic story), but between two distinct ways of attending to a focus (from-to), and between two distinct levels of existence in a stratification of levels (subsidiary-emergent). …[Polanyi]… challenges the immanent–transcendent distinction itself by recognizing the emergence of meanings and entities that transcend their subsidiary clues and material conditions. (163)

In contrast to Lowney, however, I emphasize a robust Platonism implied by Polanyi's theorizing over emergence theory.

manifestations (i.e. appearances) have far wider potential intension and extension.

When considering relations of reality/appearance or transcendence/immanence in a post-critical Platonist light, we should keep in mind, with Robjant, that "Plato's fundamental project is not...to establish an *a priori* logic with which to constrain the experience in which the world exists, but rather to understand that experience" (2012, 59). Or again, as Perl puts it, "the...point of Plato's theory...is that transcendence is not elsewhere but in our very midst. What is present in sensible things, as their properties, is transcendent form" (1999, 361). This coheres beautifully with Polanyi's post-critical project "aim[ing] to reverse the tendency of modern thought to disparage belief as merely subjective" (Mullins 2002, 211) and to overcome the "division between natural and supernatural...the phenomenal and noumenal...or an immanent world and a transcendent beyond" (Lowney 2017, 163). Indeed, the structure of comprehensive entities allows us to better understand the way that transcendent Form is present as properties in sensibles as subsidiary particulars which are indwelt and telicly harnessed toward the meaning of the entity, as glossed above.

Forms thus serve as the metaphysical basis for the logical possibility and intelligibility of reality full stop, while Primordial Forms make up the gradient of potentiality grounding the logical possibility and intelligibility of the sensible, material strata of reality. Forms are substantive in the sense that they are elemental components at the heart of reality in all its sensible manifestations. We must take them as existing necessarily insofar as all of the objects of our possible experience – from our mothers' faces, to love, to philosophical reflection, to quantum physics, to potato peelings – are inexplicable if such transcendent entities do not exist. To put it in explicitly Polanyian terms, Forms are the entities that serve as the most fundamental subsidiary particulars, and as such they bear upon all contingently possible sensible comprehensive entities. Yet, as Polanyi teaches, all subsidiary details which are tacitly indwelt can in turn (if specifiable) become the objects of our focal knowledge and thus be recognized as comprehensive entities in their own right. As we have already discussed in an earlier chapter there is no reason why a Form which is metaphysically superior in Eleatic/IFM terms cannot function ontologically and mereologically as the subsidiary particular of a (metaphysically) lesser comprehensive entity.

It may seem then that what I am suggesting amounts to a relatively straightforward foundationalist account, perhaps even a reductionism of sorts, with all of reality reducible to a set of simples – the Forms themselves. For many, and especially those unfamiliar with the intricacies of various versions of the theory of Forms, the assumption will be that Forms cannot be dismantled into a further set of subsidiary particulars making them up – that Forms *qua* Forms are indivisible, exhibiting something like a divine simplicity. This, however, is not the case. I am indeed arguing that the Forms are the metaphysical ground of all possible integrative achievement, the Polanyian and Platonic fields in which all else grows. Yet I am not arguing for an understanding of Forms that entails any sort of absolute simplicity in this vein.

This is because, as will be recalled, the joint meaning of a comprehensive entity which harnesses its set of contributing subsidiary details is not reducible to those subsidiaries. While Forms exist necessarily and are irreducible to their instantiations, they are also themselves comprehensive entities with subsidiary particulars. The subsidiary particulars include relations between the Forms as well as their potential and actual instantiations. All Forms, whether instantiated or no, include as subsidiary particulars all possible comprehensive entities that might indwell and thus instantiate them. The existence of a given Form itself is in no way dependent upon the concrete or sensible actualization of its participating particulars since, as I have argued, we must accord potency full ontological status.

A Form is actual, even if it has no instantiated particulars.[4] This is so because an uninstantiated Form exists as a comprehensive entity that teleologically unifies a set of unspecifiable subsidiaries that include its possible manifestations, actualized or not, and the actualization of any given manifestation (or all of them) does not bear upon the actualization of the Form itself, since the joint meaning is irreducible to its subsidiary

[4] Recall that the Form exists on its own, with the instance (whether potential or actual) as a subsidiary detail. The Form also exists in the instance itself as a subsidiary detail in the instance's structure as a comprehensive entity.

details. An uninstantiated Form thus has not only full ontological status, but also full metaphysical status, since it *is functioning Eleatically as a part of the field-like gradient of potential and is thus exercising its shaping power.*

Forms, then, are unities, but they are also composites. And this is no more mysterious than the simultaneous unitary and composite nature of a chunk of granite, a poem, or a child. This is, perhaps, the key metaphysical insight of Polanyi's post-critical project: to break down the unitary/composite dichotomy.

Therefore, for the post-critical Platonist metaphysics I am proposing (as for Polanyi's post-critical epistemology) there is no direct, non-circular, divinely simple, foundational justification to which all can transparently be reduced. In the process of metaphysical investigation we begin with tacit, anamnetic awareness of the Forms which allows us to engage in integrative knowing. We come to our embodied, focal awareness of the concepts grounded in the Forms through our contact with sensible particulars that participate therein, as we saw in the *Phaedo*. Once we arrive thus at the Forms and seek a better understanding of these, we use them subsidiarily and are referred immediately out to the potential manifestations of those Forms, since oscillation between focal and subsidiary awareness is, as Polanyi noted, the royal road to deepening our understanding.[5]

In case it seems to some that I am playing too fast and loose with Plato's doctrines, we ought to note that there is precedent in the tradition of Platonist interpretation for viewing the Forms as divisible in certain senses, for as Plato (1997c) himself remarks in the *Philebus*, "for any class to be alone, solitary, and unalloyed is not altogether possible" (63b). Bearing on this (as well as the often overlooked harmony between Plato and Aristotle's teachings) Frede comments that

> ...the view in the *Categories* hardly differs from the Platonic theory of forms in the *Philebus*. In the *Philebus* [15a-b, 16c-e], Plato asks how forms can be one and many. The answer is that they are at once one, many, and unlimitedly many; one, insofar as they are genus; many, insofar as the

[5] See Polanyi's "Faith and Reason" (1961). Lowney examines this integration to a Form or universal, and then the use of the Form/universal subsidiarily to understand its particulars, as a from-through-to structure, but with certain features I reject. (See Lowney 2017, 159–160.)

> genus consists of many species; and unlimitedly many, insofar as unlimitedly many things are subsumed under the various species. Here we have not only the division into genera, species, and individuals, but also the notion that species and individuals are parts of the genus or form...Platonic forms include individuals as parts. (1987, 56 f.)

Similarly, in an analysis of Plato's *Meno*, Ademollo notes that

> ...virtue is described by Socrates (78d–79e) as the "whole" (ὅλον) of which the specific virtues are "parts" (μόρια); and therefore the required definition is described as an account that "holds of it as a whole" (κατὰ ὅλου, 77a). Since this terminology has parallels inside the Platonic corpus, from the so-called "Socratic" dialogues to the "late" ones, we are entitled to regard it as an expression of Plato's genuine thinking about the forms. (2013, 43)

Ademollo goes on to list a number of places in the dialogues where Forms are treated as composite in some sense, thus lending precedent for the post-critical Platonist understanding of Forms as unitary (as comprehensive entities) and also divisible (in their own subsidiary structure and in their appearance as subsidiary details in the structure of their participating particulars):

- The form as "the same thing" present in several kinds or particulars: *Euthphr.* 5cd; *La.* 191e; *Smp.* 210b.
- Forms as "parts" of other forms: *Euthphr.* 11e-12d; *La.* 190cd, 198a; *Ti.* 30c-31a; *Sph.* 220c, 221b, 257c-e etc.; *Plt.* 262a-263b (*Sph.* and *Plt.* in the context of division by kinds).
- Forms extending "through" other forms or particulars: *La.* 192bc; *Sph.* 240a, 253d.
- The form as "one thing over all" (i.e. over all its specific kinds): *Sph.* 240a.
- The form of F as that "because of which," or "in virtue of which," or in virtue of whose "presence," other things are F: *Hp.Ma.* 288a, *Euthphr.* 6d-e, *Phd.* 100d, *Sph.* 247a. (Ibid., 43–44)

Thus

> Plato's view by the time of the *Sophist* seems to have been that being is dyadic: it is both indivisible, static unity and a divisible, dynamic whole of parts; it is both in itself and in relation. Because no being can be wholly self-enclosed and separated in complete isolation, it must have some power to affect or be affected by others. Every form is, in a way, one-in-itself, yet although each is a one, none is the one in that each one is reflected in the Platonic material principle of relative non-being or otherness. (Dombrowski 1995, 306)

I have no doubt that many of my fellow interpreters of Polanyi will, despite the forgoing arguments, harbor deep suspicions of anything that smacks of a robust metaphysical project of this nature. To them I would point out, once again, that *Polanyi's overarching purpose in working out his epistemology is to restore our ability to make such metaphysical affirmations in good faith.* I would also emphasize that in proposing his field theory, Polanyi gestures toward the necessity of the Primordial Forms – those Forms that ground the basic concepts necessary for the existence and intelligibility of sensible, material reality and which are logically presupposed by an embrace of Polanyi's epistemology with its focus on integrative noetic achievement. Polanyians affirm our ability to make contact with reality by means of such achievement, and such achievement necessarily depends on the Primordial Forms.

There is, to my mind, no need to exhaustively catalog such Primordial Forms, though discussion and debate over precisely what counts as a real property flowing therefrom is certainly no waste of time. Such a procedure, however, would have to proceed with the proper spirit of humility in the post-critical light of our position as tacit knowers. The list of Primordial Forms will never be complete, nor need it be. I have already mentioned, in non-systematic fashion, a number of candidates that I think would uncontroversially make the cut: Boundary or Limit, Equality, Pattern or Order, themselves by themselves. To reiterate a point I have made a number of times, I am convinced that positing such Forms is a metaphysical necessity for Polanyi's epistemology to explain how we make contact with reality. The ability to recognize instantiations of these

Forms is a necessary precondition for our undeniable ability to perform the skilled integrative acts of knowing that Polanyi recognized and brilliantly theorized.

As I have stated previously, the Primordial Forms must of necessity be instantiated if any sensible reality at all is to be actualized, yet it is not so with all Forms, and not even so for all Forms that, if instantiated, will manifest themselves sensibly.

In my post-critical Platonist ontology, the Primordial Forms constitute the generalized fields of potentiality that comprise the gradient sloping toward the meanings that make up sensible, material reality. They thus root intelligibility, order, and value at the beginning, both logically and temporally. But this leaves plenty of scope for many other ontological features, often called emergent, which,[6] once they acquire the creative abilities inherent in the hazardous striving towards achievement that is life, can creatively interact with reality by recognizing (and eventually naming) classes of particulars not completely subsumable as participants in only one Primordial Form (by way of examples consider humans, granite, mud, hair, frogs, redwood trees), as well as by participatively actualizing new entities, both material and immaterial, which were prefigured by possibilities inherent in the mutual relations among Forms (democracy, bicycle, the novel as literary form, etc.) by discovering these sets of relations at the level of potentiality and then actualizing them.[7] The Forms that logically prefigure the actualization of such contingent comprehensive entities are not primordial in that their instantiations were not prescribed, presupposed or preprogrammed (i.e. not necessary) within the gradient of sensible, material possibility in any explicit way.

[6] As hinted at in earlier chapters, I reject strong emergence if it is conceived of as a maneuver to avoid the logical necessity of positing the immaterial as fundamental to the material. Advocates of strong emergence hold that new ontological entities which were not prefigured by their "base" can arise from that base spontaneously. In contrast, on the basis of my development of Polanyi and Plato, I would argue that a superior mode of understanding emergence is to see it as *the process of actualization of the potential.* Since, however, both the potential and the actual have full ontological status in a post-critical Platonist metaphysics, there is no novel "coming into being" in the domain of the metaphysical existence. It is rather a shift in the modality of being. However, any helpful insight gleaned from the tradition of emergentism can be easily assimilated to the post-critical Platonist view, often without so much as modifying the terminology, so long as the difference, which lies at the level of underlying metaphysical assumptions, is kept in mind.

[7] See Phil Mullins' "Polanyi's Participative Realism" (1997).

Such Forms do not carve reality at the deepest, most fundamental joints, though they jointly help to make reality up in all its plenitude. And they are clearly real, for all the reasons Polanyi has given. To give a trite example, there are many possible worlds which include a sensible, material stratum but do not include bicycles, smartphones, people, or granite, but there are still Forms of those possibilities in those worlds.

As we saw earlier in the fourth chapter when looking at Polanyi's discussion of discovery and invention and its relation to the full ontological status we have accorded to potency, something like the Form of the Bicycle is itself a comprehensive entity, a potentially actualizable web of relations between a partially, though not exhaustively, specifiable mixture of other Forms (some primordial, like Limit, others not, like Gear) which will in part depend for its actualization as a sensible upon the actualization of those of its subsidiary particulars which, if actualized, exist as sensibles (e.g. gears, spokes, wheels). It is easy to see, then, the distinction between the necessity of instantiation of something like Limit and the contingency of Bicycle.

Thus, though the sensible instantiation of non-primordial Forms is not logically necessary, we can intuitively see that the raw, primordially substantive stuff of which they are made goes all the way down, since all of the logical and qualitative categories of which they partake are there from the logical and temporal beginning, tacitly prefigured at the level of potency and possibility, given these comprehensive entities' ontological place within the logical horizon or gradient of potentiality set by the originary field.

2 Synthesizing Plato and Aristotle Within a Post-Critical Metaphysics

Summing up the synthesis of Plato and Aristotle achieved by Middle and Neoplatonists, Gerson notes that Aristotelian thought "could be subsumed under the more capacious and ultimately true Platonic system in a way roughly analogous to the way that Newtonian mechanics can be subsumed under quantum mechanics or sentential logic can be subsumed under the predicate calculus" (2005, 4).

A move toward such a synthesis of Plato and Aristotle is appropriate both because Polanyi self-consciously modeled his post-critical project on that of St. Augustine, whose thought unfolded in the light of significant Neoplatonic influence, and also because of the sustained attention that Polanyi pays to both material and immaterial strata of reality, as we have already seen at length.

In an earlier chapter we looked briefly at some aims and commitments common to analytic neo-Aristotelianism and Polanyi's Copernican realism. After the intervening development of a post-critical metaphysics we can now see how a hylomorphic approach to Aristotelian substances is a legitimate way of understanding a subset of comprehensive entities such as living things and other material entities (I leave debates over precisely which material entities count as Aristotelian substances to card-carrying Aristotelians). For, as we have already seen, not all comprehensive entities necessarily include sensible matter as a subsidiary particular (e.g. Forms, melodic ideas, etc.). The metaphysics of comprehensive entities I have proposed as a development of Polanyi's work provides a post-critical explanation of the grounding relationship between the substantial forms in hylomorphic compounds and the Forms that make up the gradient of meaning.[8]

On this reading, then, Franklin's "semi-Platonist Aristotelianism" (2015) would be more aptly described as a semi-Aristotelian Platonism, however we won't quibble too much over names at this point, since such debate could prove distracting from the more important goal of a constructive synthesis leading to a metaphysics that is as explanatorily powerful, widely acceptable, and ontologically affirmative of the full sweep of reality as possible. The potential for rapprochement between partisans of Plato and Aristotle is rich with possibility, as we have seen already when looking at arguments proposed by Franklin and Fine. It can be seen perhaps even more clearly when physicist and neo-Aristotelian philosopher William M.R. Simpson notes that he has been

[8] A hylomorphic compound (a type of comprehensive entity) participates in the relevant Forms which are incorporated as subsidiaries that provide the ideal framework of the joint meaning, which is analogous to the substantial form in Aristotelian thought. Note, however, that – as I have argued earlier – Polanyi's post-critical approach accords the joint meaning more ontological weight given its implied interchangeability with the comprehensive entity itself.

> inclined to say that forms exist in the world, but that universals exist in the mind; they are recovered by abstraction…Do forms which are not informing some piece of matter exist? I think Aristotle would say no, but that I may have to say yes. I think of the cosmos as beginning as a single substance, which must contain the forms for other substances which emerge at later stages in cosmic history. But they are only in potency at the beginning.[9]

This sparse sketch is broadly compatible with the metaphysics I have been developing here. While Simpson's views continue to evolve (moving toward some form of trope theory)[10] the distinction between Forms and universals made in the passage above indicates that a contemporary form of the Neoplatonist synthesis is a live philosophical option outside the realms of Polanyi and Plato Studies. In parsing the ancient Neoplatonist synthesis, Gerson notes that

> when a universal is predicated of many, this is not done…in order to explain how these many can be the same; rather, the predication amounts to the recognition or acknowledgement of their sameness. Predication is without exception assumed by Aristotle to be an extra-ontological category of activity. When I say, 'Socrates is a man', what I predicate of Socrates does not explain the possibility of his being one among potentially many all with the same property, namely, being a man. What explains the truth of 'Socrates is a man' and hence the possibility of there being men is the nature or form or essence 'man' with which Socrates is somehow identified. He is not identified with a predicate. Therefore, it is at least puzzling why one should suppose that an Aristotelian theory of universals is somehow an alternative to a Platonic theory of Forms. (2004, 237)

According to Gerson, for many ancient interpreters of Plato and Aristotle, Forms are "supposed to provide an explanation for sameness in difference" (Ibid.). This sameness is then subsequently identified through the process of abstraction and crystalized in universals. He explains, for example, that the Peripatetic thinker Alexander of Aphrodisias (second – early third century C.E.)

[9] William M.R. Simpson, email message to the author, May 24, 2021.

[10] William M.R. Simpson, email message to the author, May 7, 2023.

> endorses the Aristotelian position that everything that is separate is an individual. Yet, he acknowledges that a plurality of individuals may possess the same "nature"…*This nature is neither an individual nor a universal, for the universal is just this nature only insofar as it is conceived. The universality is accidental to the nature. This nature is prior to any individual that possesses it. It is also prior to the universal.* (Ibid., 241, emphasis added)

This common nature – identified, codified, named, and tracked by the universal – is grounded in the relevant Forms in which the similar particulars participate through the process of indwelling.

Nevertheless, the process of abstraction, as we have already seen when looking at *Phaedo* and *Posterior Analytics*, does not sufficiently explain how we are able to acquire concepts and therefore requires some prior anamnetic contact with the Forms that allows for the bare recognition of content necessary for abstraction to get off the ground. We have brought this anamnetic contact under the post-critical rubric of Polanyi's theorization of the tacit and unspecifiable dimension of knowing. Furthermore, the structure of the comprehensive entity – which is intimately bound up with the theorization of the tacit and thus, due to the ontological aspect of tacit knowing, fuses the domains of meaning and being – serves here as a sort of middle term, providing an explanatory mechanism for how (via indwelling and telic harnessing) particulars participate in the Forms that define their very possibility and showing how we are able to epistemologically indwell such particulars, thus making our tacit contact with the Forms explicit *through* our grasp of the participating particulars.

Is the universal then a separate entity from the Form as Gerson, following some Neoplatonist precedent, seems to indicate? I would argue that it is indeed. It is our more or less explicit, fallible grasp of the nature of particulars which is grounded in the Forms. Universals are thus open to revision and are contingent in a way that Forms are not – they require sentient activity for their actualization and are not metaphysically prior to the particulars that give rise to them. They are, that is to say, comprehensive entities with full *ontological* status which are lower on the *metaphysical* hierarchy than Forms due to their lesser Eleatic status. After all, a confident explicit grasp of a universal *does* have a certain causal force, such as the improved ability to identify *x*. A universal does not, however, have anything like the shaping power that Forms wield. Indeed, universals in this psychological sense are an IFM generated by the Form's shaping power, which, like other instantiations of a given Form, function as subsidiary details within its (the Form's) comprehensive structure.

Perhaps this view allows everyone, Platonist and Aristotelian alike, to have their cake and eat it too (we won't worry about the nominalists for now – no cake for them). This is because the foregoing explanation shows how it might be plausibly true that at the level of actualization

a) Uninstantiated Forms can and do exist (thus appeasing Platonists), and also:
b) No uninstantiated universals exist (thus satisfying Aristotelians).

B is plausibly true given that we have distinguished universals from Forms as our focal, explicit tracking of a concept which is *grounded* in one or more Forms. Such universals thus arise from our contact with a sensible particular that is an instantiation of those same Forms. It makes perfect sense to say that the actualization of our explicit awareness of the universal is logically subsequent to our contact with an appropriate sensible particular and thus to say that a given universal's actualization requires its instantiation, which is to say that the nature tracked by the universal and grounded in the Form is instantiated. This instantiation is thus made possible by the relevant Form, which exists actually *with or without* actualized instantiations beyond the existence of the Form itself, which is actual independent of its participating particulars.

There will be, no doubt, those who object that the universal is merely epistemological and does not deserve ontological status. Such objectors will receive the reply (once again) that to exist is to mean, and that to exist at all is to be a comprehensive entity. This is all that is necessary for a place in our ontology. Such ontological standing then guarantees the possibility of mereological participation in other comprehensive entities. Functioning as a part (i.e. a subsidiary particular) within another comprehensive entity does not, however, imply a lower metaphysical status, it merely tracks a mereological relationship within an ontological structure. If this were not so, then Forms could not make ontological contact with their instantiations by showing up as subsidiary particulars within their (the instantiations') comprehensive structures, as we have previously discussed. Participation is therefore a two-way street operating metaphysically, ontologically, and mereologically whereby a particular that instantiates a Form partially indwells the Form, thus harnessing it as one of its subsidiary particulars. Yet this harnessing and indwelling is only possible because the Form, metaphysically prior and with greater Eleatic standing, has prefigured the instantiation at the level of potency as one or more of *its* (the Form's) subsidiary details in the exercise of its shaping, attractive power within the field-like gradient of meaning. Thus the potential *and* any actual instantiations are also subsidiary details within the comprehensive structure of the metaphysically prefigurative Form.

Polanyi's insight into the integrative epistemological powers that operate by toggling back and forth between tacit indwelling (attending from) and explicit focal awareness (attending to), when run through the ontological aspect and the more robust understanding of meaning as being that I have previously argued for sets the stage for a more flexible understanding of what constitutes stability in an object of knowledge since it (when paired with an understanding of comprehensive entities) allows us to recognize the ways in which beings are both unitary and composite simultaneously and able to sustain their unity even in the face of change. Furthermore, comprehensive entities are a more flexible structure than Aristotelian hylomorphic substances (as typically conceived in debates over substance) since they allow for ontological flatness *without* requiring metaphysical flatness.

Firstly, all existing things exhibit the structure of the comprehensive entity – literary plots, violins, amoebas, metaphors, etc. Following on this, all comprehensive entities are substances – metaphors can be integral parts of literary plots, for example, and amoebas of ecosystems – but not all substances are metaphysically equal. An instructive and analogous truth is familiar to all craftspeople and artists. Substances (in the everyday sense of craft) vary widely – plaster, wood, steel, wool, etc., can all be used to make things in causally effective ways. Yet this in no way means that, say, wool and steel will be equal in terms of the sorts of causes they are capable of producing or the sorts of uses to which they can be put.

What we have, then, is a stable, open metaphysical system in which the intelligible logically precedes the sensible and material, which is itself after all a particular slice of intelligibility. The stability is in part due to the flat ontology of comprehensive entities, which allows us to better understand how Forms and particulars, wholes and parts, incorporeals and corporeals, relations, etc. interact with one another. Our ability to know in a finite and partial manner other entities in this cosmos is not surprising at all, since we are a part of the cosmic system's dance. The explanatory power of this metaphysics is not of the sort desired by the critical project, however. It does not claim to render perspicuous all of the mechanisms at play without remainder, since as a post-critical proposal it takes unspecifiability and irreducibility as constitutive of reality, from top to bottom. There is therefore no non-circular justification to be found. We are thrown into the dance and must participate as best we can.

We are, as Plato, Aristotle, and Polanyi agree, materially embodied beings that come to know within corporeality, so we have a special affinity with the Primordial Forms, those which are necessarily instantiated in all possible worlds that include material elements. All use we make of the Primordial Forms – our recognition of various comprehensive entities through bare perception, our symbolic representation of such perception, our engagement in various forms of inquiry, our creation of art, etc. – *is possible only through the largely unspecifiable anamnetic medium of our tacit awareness* which connects us as living centers of achievement to the Primordial Forms. This is the grounding point for a post-critical Platonism. Thus Polanyi's insights into the fiduciary basis (the personal and tacit coefficients) of all knowing are fundamental doctrines constitutive of this approach.

Finally, a few words must be said about Polanyi's post-critical attempt to "attain a view of the world in which the universe, per se, is not 'value-free'" (*M* 161). As I read him, Polanyi's exultance in the universe's "enterprise [sweeping] …unceasingly…on toward an infinite range of higher meanings" is indicative of the classical and Platonic intuition that existence *per se* is a good (TD 90–91). But if this doesn't indicate a real property, one of the joints of reality, then I don't know what does. Thus, although I have said that my interest lies far from offering a comprehensive list of the Primordial Forms, it must be said that the most luminous of these is the Good.

References

Ademollo, Francisco. 2013. "Plato's Conception of the Forms: Some Remarks." *Universals in Ancient Philosophy.* Ed. Riccardo Chiaradonna and Gabrielle Galluzzo. Pisa: Edizioni Della Normale: 41–86.

Clark, Stephen R.L. 2015. "Platonists and Participation." *Revista Portuguesa de Filosofia.* 71/2–3. *Metafísica: Perspectivas Históricas e seus Actores*: 249–265.

Dombrowski, Daniel A. 1995. "Being is Power." *American Journal of Theology & Philosophy.* 16/3: 299–314.

Franklin, James. 2015. "Uninstantiated Properties and Semi-Platonist Aristotelianism." *Review of Metaphysics* 69: 25–45.

Fine, Gail. 2003. *Plato on Knowledge and the Forms: Selected Essays.* New York, NY: Oxford University Press.

———. 1983. "Plato and Aristotle on Form and Substance." *Proceedings of the Cambridge Philological Society, New Series* 29/209: 23–47.

Frede, M. 1987. "Individuals in Aristotle." *Essays in Ancient Philosophy.* Oxford: Oxford University Press: 49–71.

Fujisawa, Norio. 1974. "Εχειν, Μετέχειν, and Idioms of 'Paradeigmatism' in Plato's Theory of Forms." *Phronesis* 19/1: 30–58.

Gerson, Lloyd P. 2004. "What is Platonism?" *Journal of the History of Philosophy.* 43: 253–276.

———. 2005. *Aristotle and Other Platonists.* Ithaca, NY: Cornell University Press.

Grant, Iain Hamilton. 2006. *Philosophies of Nature After Schelling.* New York, NY: Continuum.

Lowney II, Charles W. 2017. "Overcoming the Scientistic Imaginary." *Charles Taylor, Michael Polanyi, and the Critique of Modernity* ed. Charles W. Lowney II. Palgrave Macmillan. 143–168.

McCabe, Mary Margaret. 1994. *Plato's Individuals.* Princeton: Princeton University Press.

Mounce, H.O. 2005. "The End of Metaphysics." *New Blackfriars* 86/1005 (September): 518–527.

Mullins, Phil. 1997. "Polanyi's Participative Realism." *Polanyiana* 6/2: https://www.kfki.hu/~cheminfo/polanyi/9702/mullins.html

———. 2002. "Peirce's Abduction and Polanyi's Tacit Knowing." *The Journal of Speculative Philosophy, New Series* 16/3: 198–224.

Murdoch, Iris. 1993. *Metaphysics as a Guide to Morals.* New York, NY: Allen Lane/Penguin.

Nagel, Thomas. 2012. *Mind and Cosmos: Why the Materialist Neo-Darwinian Conception of Nature is Almost Certainly False.* New York, NY: Oxford University Press.

Perl, Eric D. 1999. "The Presence of the Paradigm: Immanence and Transcendence in Plato's Theory of Forms." *The Review of Metaphysics* 53/2: 339–362.

Peterson, Sandra. 2008. "The *Parmenides*." *The Oxford Handbook of Plato* edited by Gail Fine. New York, NY: Oxford University Press. 383–411.

Plato. 1997a. *Parmenides*. trans. Mary Louise Gill and Paul Ryan, in *Plato: Complete Works.* Edited by John M. Cooper. Indianapolis, IN: Hackett Publishing Company.

———. 1997b. *Phaedrus* trans. Alexander Nehamas and Paul Woodruff.

———. 1997c. *Philebus* trans. Dorothy Frede.

———. 1997d. *Theaetetus* trans. M.J. Levett, rev. Miles Burnyeat.

———. 1997e. *Timaeus*. trans. Donald J. Zeyl.

Plotinus. 1968–88. *Plotinus*, 7 volumes, Greek text with English translation by A.H. Armstrong. Cambridge, MA: Loeb Classical Library.

Polanyi, Michael. 1961. "Faith and Reason." *Journal of Religion* 41/4: 237–247.

———. 1962 [1958]. *Personal Knowledge: Towards a Post-Critical Philosophy.* Chicago, IL: University of Chicago Press.

———. 1966. *The Tacit Dimension*. Chicago, IL: University of Chicago Press.

Polanyi, Michael and Prosch, Harry. 1975. *Meaning*. Chicago, IL: University of Chicago Press.

Conclusion

In summation, my project here has been to examine, develop and defend the metaphysics that undergird and are implied by Polanyi's post-critical approach to philosophy. Furthermore, I have undertaken this project in a post-critical mode, which is to say that throughout I take methodological belief to be the necessary ground upon which all rational inquiry can proceed. This acknowledgement of the fiduciary nature of all knowing, based in large part on Polanyi's theorization of the inescapable tacit and personal coefficients of knowledge, thus rejects the core dogma of the critical turn – modernity's insistence on methodological doubt as the proper basis for rational behavior, though it allows space for doubt as a valuable means of refining belief, so long as it is kept properly in check by proper acknowledgement of the inescapable, foundational nature of belief.

I have further argued that despite being known in the philosophical world as a heterodox epistemologist, Polanyi is, at root, a metaphysical thinker seeking to reorient us toward reality and to restore our ability to make metaphysical claims about non-empirical matters in good faith without thereby falling into simplistic or naïve foundationalism on the one hand nor nihilistic relativism on the other. To be sure, he seeks to do

M. E. Turkis II, *The Metaphysics of Michael Polanyi*,
https://doi.org/10.1007/978-3-031-41211-0_8

all this by epistemological means, but his epistemology is a means to a metaphysical end.

There is no doubt, however, that while Polanyi did succeed in systematizing to a significant extent his epistemology of the tacit dimension, he never developed his metaphysics with any sort of systematic rigor, a situation I attempt to remedy as my thesis proceeds by arguing that Polanyi's overall project implies and assumes a Platonism characterized by an all-encompassing metaphysical hierarchy determined by the Eleatic Principle (i.e. IFM production) (Meek 2017), a flat ontology of comprehensive entities (inclusive of immaterial and material entities), and a mereology which allows for entities which are greater metaphysically to serve as subsidiary particulars within the structure of comprehensive entities that wield less (or more) metaphysical, Eleatic power than they.

Following what I take to be Polanyi's lead, then, I have thus urged the priority of metaphysics on largely experiential, phenomenological grounds as well as on the basis of what I have called his Copernican realism whereby he, following his understanding of Copernicus's position, affirms our ability to pierce the veil of phenomena and contact reality by means of both empirical investigation and philosophy of first principles.

After tracing debates over the contours of Polanyi's realism – especially over the ontological status of the immaterial and the meaningful – I have attempted to bring Polanyi's Copernican realism into conversation with three contemporary metaphysical realisms. The first of these, Ontic Structural Realism (as argued for by James Ladyman et al.) (2007), ought to be rejected on post-critical grounds for its open scientism and its failure to allow for the inescapably fiduciary nature of the scientific process, which it takes as its metaphysical lodestar, despite some similarities to Polanyi's approach in its relatively open ontological orientation toward real patterns.

OSR's emphasis on structural relations and denial of ontological status to entities, led me to investigate a second version of realism: Object-Oriented Ontology as proposed by Graham Harman (2018). This is a more promising conversation partner. OOO, on the basis of certain insights mined from the works of Heidegger and Husserl, proposes a new theory of substance that takes the form of a flat ontology, arguing that everything that exists, from the Golden Gate Bridge, to the plot of *Moby*

Dick, to the relation known as "equality," has the same ontological status, which Harman calls the "object." I borrow Harman's understanding of a flat ontology and argue for its inclusion in a robust post-critical metaphysics while also noting some similarities between OOO's understanding of the "machine" and Polanyi's comprehensive entities. Nonetheless, I also hold that Harman's approach does not adequately explain the relationship of the sensual to the real and that further treatment of the integration of parts within "machines" could benefit from engagement with Polanyi's theories.

This leads me to look, thirdly, at the significant overlap between Polanyi's metaphysical project and that of neo-Aristotelians such as Anna Marmodoro (2013), Robert Koons (2014), and William M.R. Simpson et al. (2017), who work on the development of contemporary hylomorphism, an approach they argue has been vindicated by findings in quantum physics. My own argument is that debates over hylomorphism provide helpful conceptual tools for working out a metaphysics of comprehensive entities and that it is indeed a useful approach to looking at certain subsets of comprehensive entities, such as living things and material entities more generally. In spite of this productive overlap, I ultimately follow ancient Neoplatonist precedent as unpacked by Gerson (2005) and reject the dichotomy which pits Plato against Aristotle, seeking instead a harmonization of their positions within a more generally post-critical approach, especially with regard to Forms and universals. Nonetheless, harmonized readings of Plato and Aristotle are decidedly not the norm in philosophical debate over the last few centuries, and this has created confusion which leads, or so I argue, Polanyi himself (as well as many of his interpreters) to misunderstand Platonism and to thus miss how close Polanyi's position is in key respects to Plato's, rightly understood.

In examining the metaphysics that are explicitly present in Polanyi's corpus I show that his understanding of the ontological aspect of tacit knowing affirms the tight connection and near-interchangeability of knowing and being. As mentioned previously, Polanyi's term *comprehensive entity* – a focal whole, *Gestalt*, or joint meaning, which we grasp epistemologically by indwelling the subsidiary particulars that comprise the structure of the entity – is a critical jumping-off point for the development of a robust post-critical Platonism.

Early on in his philosophical writings (in *PK*) Polanyi primarily uses "comprehensive entity" to refer to living things and machines, later expanding its application to all objects of knowledge (from *SOM* on). Given the relationship of knowing and being affirmed by Polanyi and embodied in the ontological aspect, comprehensive entities can be taken to be the metaphysical structure of all entities or beings. I argue that this applies to the wide, pluralistic variety of IFM producing realities he recognizes, whether material or immaterial. Properly understood, then, I argue that a Polanyian ontology is flat, not hierarchical, since all existents are joint meanings comprised of the integration of their subsidiary particulars, achieved by the telic harnessing via indwelling by the comprehensive entity of those subsidiary components.

Yet Polanyi, like Plato, clearly affirms a hierarchical reality. If this hierarchy does not operate at the level of ontology, then where does it operate? My answer here is to propose the tripartite metaphysical structure sketched above: An analogical and Eleatic metaphysical hierarchy within which operates a flat ontology of comprehensive entities, with a mereological subsidiary-*Gestalt* hierarchy operative within the flat ontology.

Central to Polanyi's project as well as my post-critical Platonism is telos which is the definitive key to understanding the joint meaning of a comprehensive entity. Telos is thus on my account bound up with the intelligibility of reality, which is a fundamental aspect of the real, even in potency – for I argue that Polanyi implicitly gives full ontological status to potentialities in his treatment of stable, open systems (such as a flame) as realities pre-existing their instantiations, as do my own post-critical proposals, which make this explicit.

Thus we arrive at the conclusion that meaning is being. Borrowing terms from the hylomorphist lexicon we can say that the joint meaning of the subsidiary details is the substantial form or the principle of unity of the comprehensive entity. It acts as a formal cause as it performs its telic harnessing of its subsidiaries and a final cause as it fulfills its telos, thereby making its Eleatic impact on surrounding reality in the form of IFMs. This understanding of meaning as being helps us make sense of how comprehensive entities which are never known by reflexive sentient entities can inhere and hang together. They are comprehensive and comprehensible because they themselves comprehend and indwell their subsidiaries. In this sense, comprehension is the basis of being. For Polanyi as for Plato (and Aristotle), entities are defined from the top down, not from the bottom up.

Furthermore, all comprehensive entities qualify as substances, yet substance is not the fundamental ontological identifier since it is dependent on relation for its constitution due to the mereological relations between subsidiary particulars and the joint meanings to which they contribute. My post-critical Platonism therefore emphasizes relation over substance, but simultaneously takes an inclusive approach to substance, including in this category both the potential and the actual, the immaterial and the material, the natural and the artificial, etc.

After making the above metaphysical proposals I return to Polanyi's corpus to explore his understanding of fields and gradients of meaning. I take these to be useful and insightful analogical metaphysical postulates, despite some objections to his field theory that have been raised by other interpreters. On my account Polanyi's field theory entails a Platonist position on the existence of Forms or something very like them, though Polanyi did not himself recognize this connection. Here I begin to look at Polanyi's explicitly realist position on Universals and argue that they (Universals) are best construed within a post-critical framework as comprehensive entities. At this point, given the common understanding (or misunderstanding) of Forms and Universals as representative of competing, incompatible accounts of the phenomena of sameness in difference, I begin to use the Platonic nomenclature of Forms, though I later go on to clarify my affirmation of something like the Neoplatonist synthetic position which affirms Forms as the metaphysical ground both of the natures of particulars *as well as* universals, which are our fallible, explicit mechanism for cognitively and conceptually tracking those natures (natures, again, whose reality is shaped by and grounded in the Forms).

I have also argued that those Forms which must necessarily be instantiated for physical, sensible reality to develop at all make up a sub-class referred to as Primordial Forms. Given our own status as sensible comprehensive entities, we have a particularly close relationship with the Primordial Forms. The upshot of all this is that Forms make up Polanyi's generative fields, thus shaping, nourishing and constraining the development of reality, including sensible particulars. This account fleshes out the way that Polanyi's fields guarantee that his emergentism is reductive in Thomas Nagel's sense (2012).[1]

In connecting my proposal for a flat ontology of comprehensive entities to the Platonic ground of fields and gradients, I begin with the understanding of Forms as comprehensive entities whose subsidiary details are the particulars which participate in the Form and the relations obtaining among them. Forms as meanings thus exercise the same formal and final causal powers discussed earlier. This understanding sheds light on the way we are able to identify and track the meaningful natures that characterize the panoply of concrete particulars we encounter by means of our indwelling of those particulars in the familiar ways described by Polanyi's epistemology of tacit knowing. Crucially, this allows a space in which Aristotle's account of concept formation can unfold with more robust metaphysical grounds. Thus we can affirm not only Platonic Forms that ground the natures of both particulars and the universals that characterize our conceptual accounts of our tracking of such Form-grounded natures. This scheme also helps us conceptualize the way Forms are transcendent entities in their own right *and also* immanent in the subsidiary participative particulars. We are thus also able to see relatively clearly the way in which Forms as comprehensive entities are both universal and transcendent (in that, due to their grounding capacity, their IFMs are novel subsidiary participants as well as the universals that track the natures of those participants) *and* particular and immanent (in terms of their existence as the joint meaning of their class of subsidiaries and their mereological role as subsidiaries in the comprehensive structure of their instantiations). Finally, understanding Forms as comprehensive entities provides clues to their metaphysical structure which combines a closed, definitive nature (e.g. redness) with an open-ended, creative potentiality (e.g. IFMs consisting of novel red things).

I then examine the topic of concept acquisition in Polanyi, Plato and Aristotle under which heading we see how Polanyi's position on Universals is often taken to be roughly aligned with Aristotle's, due to a potentially

[1] "A reductive account will explain the mental character of complex organisms entirely in terms of the properties of their elementary constituents, and if…the mental cannot be reduced to the physical, this will mean that the elementary constituents of which we are composed are not merely physical." (Nagel 2012, 54).

problematic understanding of *epagōgē* as induction in the contemporary sense. I argue that Plato's more nuanced account of access to concepts through the Forms in the *Phaedo* (1997) is superior. It assumes a tacit, anamnetic awareness of the concept in question (Equality) which on a post-critical Platonist reading is explicable under the rubric of the knower's powers of tacit awareness. This tacit awareness is rendered explicit through the integrative process initiated by the knower's contact with Equality as instantiated in concrete particulars (the stones and sticks). Here we can see a harmonization of the accounts of Plato, Aristotle, and Polanyi since the Primordial Form of the Equal, exercising its Eleatic shaping power over sensible reality (and thus contributing to the field-like gradient of meaning postulated by Polanyi), sets the parameters for the necessary instantiation of Equality in material reality by means of the concrete particulars' participation in the Form (i.e. indwelling and telic harnessing of the Form – which is to say the Form's mereological positioning as a subsidiary particular of the concrete particulars). The knower's tacit awareness of the Form allows her to begin to explicitly surmise its contours in the context of her encounter with the sticks and stones.

Here is the implantation of the universal of which Aristotle speaks in *Posterior Analytics*. The universal is not the metaphysical ground of the sticks or stones' equality – it is the Form that grounds such a nature. The universal is rather the knower's fallible and explicitly-identifiable conceptual understanding of and tracking mechanism for such natures. This is not to say, however, that the universal is a non-entity. It too has full ontological status as a comprehensive entity, but it falls lower on the metaphysical hierarchy than the Form due to its relatively weak Eleatic power.

The understanding of Platonism urged here requires no Two Worlds Theory, since the relation of the transcendent to the immanent is not equivalent to the noumenal/phenomenal dichotomy problematically theorized by Kant and others. There is one world, as Perl (1999) and Mounce (2005) indicate, in which both transcendence and immanence manifest themselves in rather obvious ways. Use of the metaphorical language of "the world of the Forms" is fair enough so long as one recognizes that such language is not literal and is thus tantamount to a reference to the "'world' of the rules of chess," etc., however Polanyi's postulation of fields and gradients of meaning fulfill the same function and are, given our contemporary understanding of physical fields, more evocatively powerful since they draw attention to the permeation of the transcendent Forms throughout sensible reality as guides.

In contrast to more analytic approaches to metaphysical realism, post-critical Platonism will emphasize in both rhetoric and substance the centrality of Eros implied by the passionate personal commitment that Polanyi has shown to be constitutive of our lived experience, which is rich and integrative, but also partial and fallible *and yet* of ultimate existential import. A post-critical Platonism will thus also embrace the inescapably aporetic and analogical nature of our knowing and being, a move which along with its openness to phenomenology and its emphasis on the centrality of the personal, gives it certain affinities, broadly speaking, with Continental approaches to philosophy.

Michael Polanyi's journey as a philosopher resulted in a compelling and relatively systematic epistemology, but his motivations were always fundamentally metaphysical and ethical. It is my hope that I have been able to bring his implicit, incomplete, and unsystematic metaphysics to the fore in order to examine and further develop them in the context of an explicit awareness of the Platonism I have argued they imply. The result, I maintain, is a post-critical Platonism that hearkens back in significant ways to the Neoplatonic synthesis of Plato and Aristotle, an appropriate end given the place accorded to the Neoplatonic Augustine by Polanyi as a forerunner of Polanyi's own post-critical project.

References

Aristotle. *Posterior Analytics* trans. G.R.G. Mure. *The Internet Classics Archive*. Accessed September 24, 2019. http://classics.mit.edu/Aristotle/posterior.html.

Gerson, Lloyd P. 2005. *Aristotle and Other Platonists*. Ithaca, NY: Cornell University Press.

Harman, Graham. 2018. *Object-Oriented Ontology: A New Theory of Everything*. London: Pelican Books.

Koons, Robert C. 2014. "Staunch vs. Faint-Hearted Hylomorphism: Toward an Account of Composition." *Res Philosophica* 91/2: 151–177.

Ladyman, James, Ross, Don, and Collier, John G. 2007. *Every Thing Must Go: Metaphysics Naturalized.* Oxford: Oxford University Press.

Marmodoro, Anna. 2013. "Aristotle's Hylomorphism without Reconditioning." *Philosophical Inquiry* 37/1–2: 5–22.

Meek, Esther Lightcap. 2017. *Contact with Reality: Michael Polanyi's Realism and Why It Matters.* Eugene, OR: Cascade Books.

Mounce, H.O. 2005. "The End of Metaphysics." *New Blackfriars* 86/1005 (September): 518–527.

Nagel, Thomas. 2012. *Mind and Cosmos: Why the Materialist Neo-Darwinian Conception of Nature is Almost Certainly False.* New York, NY: Oxford University Press.

Perl, Eric D. 1999. "The Presence of the Paradigm: Immanence and Transcendence in Plato's Theory of Forms." *The Review of Metaphysics* 53/2: 339–362.

Plato. 1997. *Phaedo* trans. G.M.A. Grube, in *Plato: Complete Works.* Edited by John M. Cooper. Indianapolis, IN: Hackett Publishing Company.

Polanyi, Michael. 1962 [1958]. *Personal Knowledge: Towards a Post-Critical Philosophy.* Chicago, IL: University of Chicago Press.

———. 1966. *The Tacit Dimension.* Chicago, IL: University of Chicago Press.

———. 2013 [1959]. *The Study of Man.* New York, NY: Routledge.

Simpson, William M.R., Koons, Robert, and Teh, Nicholas J. 2017. *Neo-Aristotelian Perspectives on Contemporary Science.* New York: Routledge.

Index[1]

[1] Note: Page numbers followed by 'n' refer to notes.

M. E. Turkis II, *The Metaphysics of Michael Polanyi*,
https://doi.org/10.1007/978-3-031-41211-0

The manufacturer's authorised representative in the EU is Springer Nature Customer Service Centre GmbH, Europaplatz 3, 69115 Heidelberg, Germany. If you have any concerns regarding our products, please contact ProductSafety@springernature.com

Printed and bound by CPI Group (UK) Ltd, Croydon, CR0 4YY

01/10/2026

02225894-0004